Wildfires in America

Recent Titles in the
CONTEMPORARY WORLD ISSUES
Series

Women and Girls in STEM Fields
Heather Burns Page

Remote Learning and Distance Education
William H. Pruden III

Religious Belief and Science
Glenn H. Utter

Child Welfare in America
Yvonne Vissing

Water in the West
Jacqueline Vaughn

Teen Runaways in America
Michele Wakin

Electronic Waste
Josh Lepawsky

Youth Activism
Hava Rachel Gordon

Extreme Weather and Climate Change
Mariangelica Groves

Black Lives Matter
Shaonta' Allen, Simone N. Durham, and Angela Jones

Human Rights around the World
Yvonne Vissing

Political Representation in America
Lori Cox Han and Caroline Heldman

Paying College Athletes
Jimmy Sanderson, Editor

Privilege in America
Kelly L. McFaden and Sheri C. Hardee

Wildfires in America

A Reference Handbook

Jacqueline Vaughn

BLOOMSBURY ACADEMIC
NEW YORK • LONDON • OXFORD • NEW DELHI • SYDNEY

BLOOMSBURY ACADEMIC
Bloomsbury Publishing Inc, 1359 Broadway, New York, NY 10018, USA
Bloomsbury Publishing Plc, 50 Bedford Square, London, WC1B 3DP, UK
Bloomsbury Publishing Ireland, 29 Earlsfort Terrace, Dublin 2, D02 AY28, Ireland

BLOOMSBURY, BLOOMSBURY ACADEMIC and the Diana logo are trademarks of Bloomsbury Publishing Plc

First published in the United States of America 2026

Cover image: California Wild Fire Dimple Bhati/Getty Images

Library of Congress Cataloging-in-Publication Data

Names: Vaughn, Jacqueline author
Title: Wildfires in America : a reference handbook / Jacqueline Vaughn.
Description: New York, NY : Bloomsbury Academic, 2026. |
Series: Contemporary world issues |
Includes bibliographical references and index.
Identifiers: LCCN 2025032919 (print) | LCCN 2025032920 (ebook) | ISBN 9798765131367 hb | ISBN 9798765131381 epub | ISBN 9798765131398 adobe pdf
Subjects: LCSH: Wildfires–United States–Handbooks, manuals, etc. | Wildfires–Economic aspects–United States | Forest fires–Environmental aspects | Climatic changes–United States
Classification: LCC SD421.3 .V38 2026 (print) | LCC SD421.3 (ebook) | DDC 363.37/90973–dc23/eng/20250922
LC record available at https://lccn.loc.gov/2025032919
LC ebook record available at https://lccn.loc.gov/2025032920

ISBN: HB: 979-8-7651-3136-7
ePDF: 979-8-7651-3139-8
eBook: 979-8-7651-3138-1

Series: Contemporary World Issues

Typeset by Integra Software Services Pvt. Ltd.
Printed and bound in the United States of America

For product safety related questions contact productsafety@bloomsbury.com.

To find out more about our authors and books visit www.bloomsbury.com and sign up for our newsletters.

Contents

Preface

It was an eerie comparison in early January 2025. Residents living east of the Rocky Mountains were shoveling snow and wearing puffer coats as winter hit bringing fierce winds and close to zero temperatures. In Southern California, hundreds of thousands of residents were being evacuated from their homes as unseasonably warm temperatures, Santa Ana winds, and drought combined to fuel fires that killed thirty people, burned thousands of acres, and destroyed thousands more structures.

These contrasting scenes may be the new normal. The old predictions about fire weather are being replaced with projections of how climate change will increase the median burned area in many parts of the United States. The timeliness of this book is not a result of the 2025 fires or the new administration in the White House. It is, instead, a reflection of a combination of fast-moving developments in science, public policy, the political sphere, and nature.

Chapter 1 begins by summarizing how humans have used fire from early prehistory through contemporary attempts to use but also manage wildfires. A brief explanation of indigenous people and the role of fire in early American cultures then moves into one of the more contentious issues, light burning, or as it is now called, prescribed burning. The chapter identifies the types of wildfires, and how they are caused primarily by humans. The second half of the chapter covers wildfire management from suppression and exclusion to attempts to improve forest health and the agencies and organizations involved.

Chapter 2 connects causes to impacts, showing how wildfires have an effect on what it costs to fight a fire, as well as consequences on human health, the environment, and the economy. Because approaches are changing, the chapter moves into the area of wildfire mitigation—what works and what has not worked—and the issues policymakers face in what historian Stephen Pyne calls the Pyrocene era.

In order to provide an inclusive perspective on wildfires, Chapter 3 is a collection of what those who work in the field believe are key issues, solutions, and approaches to dealing with wildfire. These essays have been chosen for their unique take on the problems we now face in management and fire science.

Chapter 4, in contrast, provides profiles of the people who have had an enduring effect on what we know about wildfires and how to deal with them.

The last four sections of this book are designed to provide additional information and resources. Primary documents, including government reports, statutory material, and key data expressed as tables, make up Chapter 5. In Chapter 6, an exhaustive bibliography of not only the sources used in the book, but also other important books, articles, and media has been created for those seeking further investigation. In order to put events in perspective, Chapter 7 is a lengthy chronology that helps the reader develop a timeline of not only significant wildfires, but also legislation and key policies. Since fire science and policymaking have a language not always understood by those outside the professions involved, the last section of the book is a short, easy-to-understand glossary of key terms and phrases.

This book is dedicated to my husband, Bob Easton, whose support has never faltered during the challenges he has faced with Parkinson's disease.

PREVENT
FOREST FIRES
IT PAYS!

1

Background and History

Introduction: The "Ancient Alliance" between Fire and Humanity

When did humans "discover" fire? There are cartoons that picture people living in caves somehow capturing fire resulting from a bolt of lightning. However, the timeframe when fire was first "controlled" by humans more likely dates back as far as 1.9 million years ago, to the time of *homo erectus*. Neanderthals may have been able to control fire about 400,000 years ago, and fire use may have occurred in Africa during the Lower Paleolithic period, also known as the Early Stone Age, which spans from about 3.3 million years ago to 300,000 years ago. Early humans, according to archaeologists, used fire for warmth, as a source of light when they lived in dark caves, and eventually for cooking or what has been called anthropogenic fire. Eventually, fire became a means for conducting slash and burn agriculture and, then, a way to clear the land and influence the distribution of species and the structure of ecosystems, otherwise known as biotic forcing. The Anthropocene, although not a precise unit of geologic time, began when humans used fire to shape their landscape to make their world a more habitable place, turning the wildland fire into agricultural fire. Historian Stephen Pyne notes that the more contemporary term is the Pyrocene—the age in which humans manipulate fire, as we do today, calling wildfire among the oldest of natural phenomena. He says wildfire is both inevitable and essential, a cultural phenomenon, and among man's oldest tools (Pyne, 2023).

A group of Civilian Conservation Corps enrollees seated aboard a US Forest Service fire truck on the way to New Hampshire's White Mountain National Forest, May 1939. Part of a federal work relief program, CCC firefighters became an easily mobilized force to fight fires, build fire lookout towers and roads, construct water basins and ponds, remove flammable material, and create fire breaks (National Archives).

What did wildfire look like prior to European exploration and settlement? Research conclusions vary. There is a significant amount of evidence that much of North America experienced historically infrequent (meaning every 200 to 800 years), preindustrial high-severity wildfires on a large scale before European exploration. Aside from the natural ignition source—lightning—fire probably came to North America with some of the original travelers who brought it from Asia, then with explorers and conquerors from Europe. There is an entire body of study on how American Indians used fire to change ecosystems through intentional burning. Although prehistoric fires were once presumed to have been started by lightning, there is no systematic record of fire events. Gradually, that view has changed, as tree ring research, detailed late nineteenth-century land surveys, and the examination of fire scars have made clear that some fires were intentionally set.

Indigenous Peoples' Beliefs and Use of Fire

Long before anthropologists and other researchers studied the role of fire among Native Americans, much of what was known about indigenous people's culture related to fire came from explorers and traders. The journals of the Lewis and Clark expedition, for example, relate how in 1806, in Idaho County, Idaho, the native people entertained the crew one night by setting fir trees on fire, telling the men that they did so to bring fair weather for their journey. In 1826, British botanist David Douglas found that the Paiute tribe in Oregon used fire to encourage deer to graze on vegetated areas where they could be killed more easily than in forested areas. Canadian anthropologist Henry T. Lewis was one of the first researchers to question the traditional "Garden of Eden" view of indigenous people and the environment by exploring their technologically sophisticated fire regimes. In his books and articles, he concluded that there were at least seventy different reasons why native people used fire to manage the land.

Modern researchers use far fewer explanations, ranging from hunting and crop management to warfare and signaling. Other studies have concluded that Native Americans set fires for different purposes at certain times of the year. In northern North America, fires were set in the early spring to clear the land for growing crops and stimulating berry growth. Native people also set fire at regular intervals, sometimes as often as every five years. That frequency meant that there was less fuel to feed the fires, and fires were less intense and less likely to become ground fires than crown fires.

Some tribes, such as those in the northern Great Plains, lit large fires to burn forests and prairies so that bison would be attracted to lush grasses where they could be hunted. Fires were even used to herd grasshoppers as a food source. In the Black Hills of South Dakota, the Lakota tribe burned the vegetation where they built their campsites, protecting them from wildfires and from enemies hiding in the dense brush. Scientists studying the Salish and Kootenai tribes in Montana and Idaho determined that they used fire to attract game, enhance the growth of berries, make travel easier, and communicate with other tribes. In the eastern United States, tribes used targeted burning to promote the growth of oak and chestnut trees to achieve desired crops.

Native Americans' practices related to fire began to disappear, however, due to European influences and beliefs. In California, the development of the mission system of religious outposts and the colonization of indigenous peoples suppressed native burning practices, causing many grasslands to be replaced by shrublands. In contrast, the culture of woodburning and even broadcast burning in the South that had been practiced by local tribes in the region's pine woods were used and adapted by European settlers for agricultural purposes. By the 1880s, industrial logging appeared, and for decades, the South was the center of that economic sector until the industry collapsed. The concept of fire exclusion that had been promoted in the early 1900s became policy well into the next century, despite the occasional conflagration like one in 1898 in North Carolina that burned over 3 million acres.

What we know now about how native peoples used fire comes from oral tradition, tree ring and fire scar research, travelers' and explorers' journals, and a growing canon of research literature. But there are also drawbacks in trying to explain how indigenous tribes in this country used fire. Researchers attempting to do so often rely on sources that may be unreliable, based on hearsay or misinterpretation. They may use overgeneralizations that treat all regions or tribes alike when there are differences among them. Others attribute Indian use of fire to landscape-scale change when in fact there are natural causes and explanations. Those are just a few of the reasons why it is advisable to tread lightly when relying on accounts that are 400 years old, or when imprecise or unfamiliar languages are involved. The term "cultural burning" has been increasingly used to describe the practice of using intentionally lit controlled fires for a cultural purpose, such as promoting the health of vegetation and animals that provide food, clothing, and ceremonial items. Some anthropologists have described the practice as a way in which tribes returned to their philosophy of fire as medicine.

Although there has been a significant amount of study of native practices in the West, for many years, the Eastern half of the country was believed to have been devoid of fire use by aboriginal people. Turn-of-the-century studies led to the conclusion that it was "inconceivable" or even "preposterous" that native people used fire regularly or on a large scale in the region. As researcher Gordon M. Day has noted, the size of the Indian population has not been and may never be determined accurately. He believed that the people in the East used fire to clear land for their villages, and that the villages were periodically relocated to expand a tribe's influence (Day, 1998). There is also evidence that even isolated tribes in the region used fire. The Wampanoag of Martha's Vineyard, off the coast of Massachusetts, for instance, were known to have used large-scale burning to open up the island's landscape to make it easier for them to hunt game and to help them maintain berry patches and cultivate crops.

Those practices may have been copied by the Pilgrims after their arrival on nearby Cape Cod. In the 1620s, Thomas Morton was said to observe that burning allowed the "savages" [*sic*] to pass through the countryside on beaten paths, according to C. Keith Wilbur, who studied the eighteen pre-colonial tribes of the region. Although Wilbur notes that the region was inhabited as long as 32,000 years ago, critics argue that there is no evidence that humans inhabited New England earlier than 15,000 years ago, following the end of the Wisconsin Ice Age (Wilbur, 1996).

The use of fire by indigenous people in the West, about whom more is generally known, is based on more contemporaneous accounts, some of which came from the Lewis and Clark expedition between 1804 and 1806. Their journals as they traveled through the Louisiana Purchase territory often provide very specific examples of native life, including how fire was used to clear the land, and descriptions of what was believed to be unspoiled, primeval wilderness. Aside from those types of records, the pictures that are painted of indigenous fire use portray a variety of practices based on vegetation and culture. There are several accounts from South Dakota's Black Hills where the Lakota tribe burned prairie vegetation to deprive bison of their winter forage while, at the same time, improving grazing for their horses.

The picture is different in the Pacific Northwest, however, where indigenous tribes found their food in the streams and ocean, not through cultivation, so fire was used less often. Alaska's natives varied in how fire was viewed, with some, such as the Koyukon in the western interior, considering fire as a destructive force. In comparison, the Gwich'in in the eastern interior actively used fire to help them pursue game and create fencing. Little is known about pre-settlement fire use in some regions, such as Hawaii, but there are numerous accounts of how tribes weaponized fire, especially on the plains where grass ranges were burned to drive away enemies. In the Carolinas, fire was used to drive deer into areas where they could be more easily killed, and in Texas, fire was used to kill mosquitos and to get lizards to come out of the soil so they could be eaten. Thomas Jefferson responded to an inquiry from John Adams about the prevalence of wide swaths of prairie in the West by suggesting they were the result of "fire hunting." By the 1840s, most of the West was still sparsely settled, and little was known about indigenous fire practices until the late nineteenth and early twentieth centuries that anthropologists like Franz Boa, Frank Speck, and Louis Kroeber sought to learn more about the cultures of Indian people.

There is agreement among most researchers, though, that the use of fire by indigenous people is a complex mosaic of practices affected by external influences and constraints. Native people and their relationship to the land changed as they migrated, as they encountered settlers and conquerors, and as they were displaced. John Wesley Powell, who had an intense personal interest in native tribes and later became the director of both the US Geological Survey and the Bureau of Ethnology, concluded that in order to increase the extent and value of the forests, the fires could be curtailed by the removal of the Indians. The consequences of the loss of indigenous fire and cultural insensitivity toward native practices would affect the next phase of wildfire management.

The Light Burning Debate

The use of fire in early American settlements often followed the traditions and practices of the indigenous people who lived on the land when explorers and colonists arrived. For the most part, they used wildland fire to open up the landscape for pastures, agriculture, and eventually, to harvest timber. But later on, settlers adhered to the belief that fire was a destructive force that needed to be controlled. The native peoples' frequent practice of using fire to reduce the buildup of dangerous vegetation and other fuel that could become a deadly wildfire—called "light burning"—was termed Paiute forestry as a derogatory reference to the Paiute Indians of California and Nevada.

During his studies of Native Americans and the use of fire to manage vegetation over the last 2,000 years, retired Pennsylvania State University researcher Marc Abrams found that prior to European settlement, even with their low population, tribes were able to repeatedly burn large areas of the Eastern United States. He says they were excellent vegetation managers, and that the indigenous people understood the need to burn the forest understory regularly to regenerate the plant species they wanted for food and to feed the game animals they relied on. When settlers moved in and human-managed burning was curtailed, the composition of the forests changed, reducing fire-adapted species such as oak, hickory, and pine (Nowacki and Abrams, 2008). US Department of Agriculture Division of Forestry scientist Franklin Hough valued "broadcast burning" as did John Wesley Powell after his work with the Paiute. Timber owners routinely used light burning in the pineland forests of the South, and Frederic Clements, an early ecologist before the term was used, advocated the use of fire to manage high-elevation lodgepole forests in Colorado.

One of those involved in the light burning controversy was Gifford Pinchot, the first chief of the US Forest Service, who had studied the role of fire in Douglas Fir forests. He noted that the species appeared to reproduce only in areas that had burned. But Pinchot firmly believed that fire should be kept out of western forests in order to protect valuable timber resources as part of the agency's organizational mission under the 1897 Organic Administration Act. It was in northern California's pine forests that light burning became a recognized practice by the turn of the century. Timberland owners adapted "the Indian way" to reduce the fuel that accumulated in mature forests, and ranchers saw it as a way of opening up pastures for their cattle and sheep. The so-called protectionists sought to eliminate fire entirely, while another California faction sought to make light burning mandatory. The millions of acres burned by the fires of 1910, however, seemed to settle the controversy even though the US Forest Service was facing an internal debate between light burning and systematic protection. Despite years of experiments to test the viability of light burning, studies of its economic feasibility, committees, and conferences, light burning was neither extirpated nor endorsed. Fire protection, covered later in this chapter, evolved slowly after the turn of the century, following the ebb and flow of light burning as a wildland fire practice.

Types of Wildfires

Wildfires are generally classified into three categories, although there are differences based on the type of fuel and the intensity of a fire. Surface fires are ones that burn rapidly but at a low intensity. They may burn light fuels, such as vegetation, but they do not usually harm mature trees or their root systems. Some types of vegetation are more likely to burn than others, and climate conditions, such as drought, may influence the behavior of surface fires. Ground fires, in comparison, are the ones that burn with the most intensity, creating their own "fire weather" and destroying all vegetation down to the soil level. These types of fires are also the most infrequent, however. Crown fires usually start from ground fires, and as the name suggests, they occur in the upper sections of trees. This type of fire can result in even more damage when the branches of the tree fall to the ground, further spreading the fire. In addition, crown fires are responsible for spreading embers that can be blown miles from the main fire, igniting new fires and causing even more damage.

Some now include the term "fast fires" in the mix to describe those that thrust embers into the air ahead of encroaching flames. A 2024 study collaboratively produced by several researchers from multiple institutions in the West determined that 90 percent of the destruction from the 60,000 blazes that took place between 2001 and 2020 in the contiguous United States occurred during fast fires, even though they represented less than 3 percent of the blazes on record. The images of the Pacific Palisades and Eaton Fires in California in 2025 vividly illustrated the role that winds of up to 100 miles per hour played in unleashing the embers that made widespread destruction inevitable in what was called a perfect storm event.

Wildfire by the Numbers

The media, researchers, public officials, and members of the public want increasingly more information about wildfires—how big they are, how much damage has occurred, whether anyone has been injured or killed, and when they will be contained. The answers depend on which source they turn to, as well as when the information was compiled. In the 2025 fires in southern California, misinformation and disinformation were key factors in determining how the wildfires and agency responses would be framed.

One reputable source for learning about wildfires that occur on federal lands is the National Interagency Fire Center (NIFC), which compiles the Wildland Fire Incident Locations dataset. The NIFC also tracks the number of fires and acreage burned using reports generated by local, state, and federal agencies. Another source is the non-profit National Fire Protection Association, which administers fire safety standards.

There may be inaccuracies in historical data because reporting was often inflated or sometimes underestimated. In addition, prior to 1983, federal wildland fire agencies did not track wildfire information in the same way as is done now, so figures before that year may be left off of tallies. Similarly, there are specialized distinctions used by different agencies that may make comparisons difficult. The National Interagency Coordination Center (NICC), for example, refers to ten geographic areas in many of its tables, rather than individual states, and the data on human-caused wildfires captures statistics for nine regions plus the state of Alaska. Information on fires caused by lightning and those caused by humans were not collected by NICC until 2001.

The phrase "worst wildfire" that is often used in the media is misleading and really does not reflect the complexity of today's wildfire losses. For instance, most of the deadliest fires happened over 100 years ago, when our knowledge about fire behavior and fire research were just beginning. Firefighting was a new occupation and resources were very limited, so putting out fires was much more difficult. A phrase such as "large fire" has a specific meaning, referring to one that damages over 100 acres of timber or 300 acres of grassland. Some sources combine different types of fires, reporting both structure or property damage, and acreage or resource losses, such as the value of timber burned in a national forest.

Nationwide, the Congressional Research Service has reported that the number of acres burned annually is an extremely variable figure, although generally there has been an increasing trend. Since 2000, the acreage figure is more than double the annual acreage burned in the 1990s, although a greater number of fires burned annually in the 1990s (Congressional Research Service, 2023). It is still difficult, however, to draw national-scale conclusions about wildfires.

Analyzing wildfire trends on the basis of how many acres burn is not necessarily an accurate way of determining severity or what constitutes the worst wildfires. In some instances, management practices have changed as research indicated that wildfires that were allowed to burn reduced accumulated fuels and supported ecosystem health, particularly in remote areas of forest. For instance, the fires that burned across Yellowstone National Park in the summer of 1988 actually consisted of eight large fires and dozens of smaller ones, becoming "complex" fires. Together, they burned about 683,000 of the park's 2.2 million acres. One of them, the Fan Fire, was naturally caused and was initially allowed to burn. The combined fires had burned about 17,000 acres when extreme fire behavior caused park officials to decide to suppress all new and existing fires.

Wildfires are often thought of in terms of how much they cost, but those numbers are inconsistently applied. Wildfires are often considered expensive because of the cost of the structures burned, usually based on insurance losses. For instance, the November 2018 Camp Fire in California had an estimated dollar loss of $10 billion, largely because of the number of structures lost—nearly 19,000. That figure would be about $24 billion in today's value using the Consumer Price Index inflation calculator from the Bureau of Labor Statistics. One law firm stated that the Camp Fire losses should also include the estimated $4 billion in uninsured losses. In comparison, the October 1993 Laguna Beach Fire in California destroyed 441 structures at an estimated cost in 2024 dollars of $1.1 billion. Damage costs are generally higher in more recent years because the country is more built up than in the previous decades, with more expensive buildings that are then rebuilt. Adding on the costs of fire suppression can also be considered part of the equation, as is discussed later in this chapter.

Certainly, loss of life can be considered part of a wildfire's cost, although the science of putting a price on a person's death is still controversial. The largest number of recorded fatalities resulting from an American wildfire is 1,152, the number of lives believed lost in Wisconsin's October 1871 Peshtigo Fire. The next highest fatality count, at 453 persons, is from the combined Cloquet and Moose Lake Fires in Minnesota and Wisconsin in October 1918.

Some wildfires can have additional costs that may be difficult to quantify. The 2024 Smokehouse Creek Fire in Texas, which became the largest fire in the state's history, burned for three weeks over sparsely populated rural land, killing two people and resulting in the deaths of thousands of cattle and other animals. Assessing the costs of this type of fire can differ significantly than estimating wildfire costs related to structural losses. One type of dollar loss does not necessarily carry the same weight as another type.

Long before the 2025 wildfires in Los Angeles, historian Stephen Pyne pointed out in his book *World Fire* (1995) that the Southwest is considered "the most interesting fire province in America. In the realm of wildland fire, the Southwest is always close to the center, always the partner of whatever region may lead the dance." The introduction of sheep and cattle completely altered and restructured the regional fuel complex, with cacti replacing grasses, transforming surface fires into crown fires. Tree density and fuel load exploded, creating a forest made up of dog hair thickets. Biodiversity declined, and as Pyne writes, domesticated fire was replaced by feral fire. California is usually considered the epicenter of the country's most visible, if not its most destructive, wildfires, beginning with the widely televised Bel-Air fire in 1961, the Mandeville Canyon Fire of 1978, and the 1991 Tunnel Fire in the hills above Oakland and Berkeley (Pyne, 1995).

Other parts of the country are not immune to the devastation of wildfires, however. In the eastern portion of the United States, fires in Maine in 1947 portended what other parts of the country would soon experience; as people began to build houses in dense patterns and closer to undeveloped lands, wildfires began to occur in the wildland-urban interface (WUI). Texas ranks as the state with the most wildfires in some years, and Alaska often has the most acres burned, although fewer structures are involved and there is generally less loss of life. More wildfire ignitions occur in the East than in western states, but wildfires in those regions tend to be smaller in terms of acreage burned.

Critics of existing measures of severity assessment, such as Sarah Hayden (2024), question whether the data in interagency reports such as Monitoring Trends in Burn Severity (MTBS) overstate the wildfire problem. She notes that the 2024 Tanques Fire in the Santa Fe National Forest in New Mexico, which started as a lightning strike, covered only 13 acres in its first week. According to the US Forest Service (USFS), the fire was managed under a confine and contain strategy. At that point, the USFS decided to expand the fire up to a planned perimeter of 7,000 acres, more than 500 times the original size, using ground firing and aerial operations, to improve wildlife habitat. It is hard to measure, she says, which part of the "wildfire" was due to intentional burning and which was "natural" wildfire. But it is clear, Hayden notes, that the majority of the acres burned were due to the Forest Service's own ignitions. The USFS called it a wildfire, though the agency does not differentiate in its published data between planned ignitions during wildfire management operations and fire from an initial natural ignition. The bottom line is that there is no way to get an accurate picture of wildfire trends and costs on a small scale, making comparisons problematic (Hayden, 2024).

Fire Weather and Drought

The nation has shown an increased interest in wildfires since 1980, a point in time when temperatures rose and the climate (and subsequently, vegetation) became drier, and when fire suppression costs grew. Although the scale and scope of those factors are subject to discussion and debate among fire professionals and researchers, there is some consensus that trends indicate fires have become fewer in number but larger in intensity. The debate is complicated by general agreement that fires are a natural process but disagreement as to the role of climate change and how to reduce losses by preventing wildfires.

Californians, for instance, are accustomed to periods of high winds that signal the potential for wildfires. Even news reporters refer to "fire weather" caused by the Diablo winds in the San Francisco Bay Area or the Santa Ana winds in the southern part of the state. In the past, reports referred to the West's fire season as June through November or early December. But there is nothing typical about what has been happening more recently. One study of fires in California that were at least 15 square miles in size found that since 1984, of the 423 fires in the sample, just six wildfires burned more than 2 square miles in any January. The largest had been the Viejas Fire in the mountains east of San Diego in 2001, which burned 17 square miles. In late December 2021, Colorado's Marshall Fire became the most destructive wildfire in the state's history, turning a grass fire into a conflagration that burned 1,084 homes in just two days. "Winter fires should be an oxymoron," said one researcher, while another observer noted "There's no fire season, fire year. It's year round."

Even before 2025 began, fire forecasters were issuing warnings that conditions in Southern California were ripe for massive infernos. The National Interagency Fire Center (NIFC) issued a bulletin on January 2, 2025, forecasting "above normal significant fire potential" even though the region typically experiences rainfall throughout the winter months. Temperatures, the NIFC noted, were "an impressive two to six degrees [Fahrenheit] above normal in December, allowing vegetation like grasses and chaparral to dry out—ideal fuel for wildfires." The National Weather Service issued its highest level of alert for much of Los Angeles County, saying wind gusts could reach 100 miles per hour in some areas, producing "extreme fire behavior." The red flag alert was the first in January since 2021.

An extreme weather pattern, termed "weather whiplash," caused the region to have drought from 2020 to 2022, then record rainfall in 2023 and early 2024, followed by some of the hottest ever recorded temperatures. Precipitation in Southern California was about 2 percent of normal by early January 2025, and much of the state had not seen rain for months on end. In addition, the record highs of 2023–4 triggered underwater heat waves that can bend the jet stream—a ribbon of fast-moving air at high altitudes that affects the weather below, creating pockets of warm air under high pressure that circulates in a clockwise pattern. These conditions allowed air from a high-pressure system over the Great Basin northeast of Los Angeles to flow over inland mountains, downhill toward the coast of California. As the air moves to lower altitudes, it compresses and squeezes through canyons, heating and drying out, gaining in wind speed and becoming turbulent and unpredictable. This was the fire weather that greeted residents of the LA Basin in January 2025.

The amount of plant growth in the state in January 2025 was nearly double what is typical for California. The desiccated, flammable grasslands had been dried out by the record-setting hot temperatures of 2024, although in June, only 1 percent of the state was considered abnormally dry according to the US Drought Monitor, which evaluates drought conditions across the country. There were 6,078 fires in California in the summer of 2024, burning three times as much total acreage as in 2023, nearly 1 million acres total.

The fire behavior in the California blazes echoed what was going on with the weather. Both during and after the 2025 fires in Southern California, officials and researchers questioned whether the wildfires were actually urban fires that spread into the wildland-urban interface by embers that traveled 2 to 3 miles, driven by hurricane force winds. They point to the Eaton Fire's random ignition points across the city of Altadena rather than an organized spreading flame front. The destruction was unsystematic, torching everything at one house while leaving another completely untouched. When there is a scattered pattern of burned homes in a high-density development, showers of burning embers cross streets and landscapes unpredictably.

With the California blazes, firefighters encountered extreme conditions that even seasoned fire officials did not envision. Extreme fire behavior included changing winds that created what is termed a particularly dangerous situation (PDS). Under typical conditions, Santa Ana winds cause fires to burn through canyons in a long and narrow corridor pattern, but in this case, the fires burned laterally, often switching course throughout the day. In the Altadena area, mountain wave winds occurred, a pattern where gusts rapidly drop down mountain slopes, gaining strength when they hit flat ground, reaching 80 to 100 miles per hour. The gusts then carry embers that sometimes invade structures through small openings, burning from the inside out. These types of fires are not spread by burning vegetation but by air.

Long-range forecasters from AccuWeather predicted that between 7 and 9 million acres would burn across the United States in 2025, figures near to above the historical average. They expected that 60,000 to 75,000 fires would ignite over the course of the year, of which 7,500 to 9,000 would take place in California. The company's senior meteorologist noted that despite a few early spring storms, drought conditions were deeply entrenched. Abnormally dry conditions paired with warm weather could also boost the fire danger across the Gulf Coast states and along the Eastern seaboard. In March 2025, wildfires broke out in areas like North Carolina where Hurricane Helene had downed large swaths of forests, dramatically increasing the amount of fuel available for fires. In states like Georgia and Florida, periods of dry weather with below-normal thunderstorm activity can lead to temporary flare-ups of blazes (Lada, 2025).

Human-Caused Wildfires

One of the more revealing statistics about wildfires is how they start. According to the NIFC, from 2013 to 2023, human-caused fires accounted for 88 percent of all wildfires in the United States. Equally important to understanding wildfires is the general cause categories developed by the National Wildfire Coordinating Group (NWCG), which provides common data standards used by fire agencies. These categories of human-caused wildfires are important in developing fire prevention strategies, for prosecuting offenders, and in preparing pre-suppression strategic plans.

Arson

Defined as the deliberate or malicious setting of fire with the intent to damage or defraud, arson accounts for as few as 20 percent of human-caused fires, and as many as 70 percent, depending on the jurisdiction. The difference between the two statistics is sometimes because a fire is classified under one cause, or remains undetermined, until further investigation reveals a pattern of activity.

Kentucky, for instance, has a particularly high rate of arson-caused fires. While 99 percent of fires in the state are human-caused, nearly 60 percent are attributed to arsonists. Although the NIFC tracks arson, there is no central database that tracks information about the type of perpetrator involved.

Wildfires started by young men, such as the Jones Road Wildfire in Ocean County, New Jersey in April 2025, have received a considerable amount of media attention. In this case, a nineteen-year-old and seventeen-year-old were arrested after setting pallets on fire using gasoline in the Forked River Mountains Wilderness Area. The fire burned about 15,300 acres, with one commercial building completely destroyed, along with multiple outbuildings and vehicles. When the wildfire crossed the Garden State Parkway, 5,000 residents were evacuated and several roads were closed. While incidents like this one do not always cause much damage, the criminal charges can be the same, including aggravated arson.

Fires caused by arsonists often increase after a natural disaster, with the perpetrators believing it represents an opportunity to commit the crime when law enforcement officers are stretched thin and there is less chance they will be caught. Arson can also be a copycat crime, driven by a desire to get attention. Mental health issues can be another reason why offenders commit arson, with some feeling marginalized or disconnected from society, anger, or even a desire for revenge. In January 2025, eight persons were arrested after setting small fires just after the Los Angeles blazes. One woman set piles of trash on fire, admitting she enjoyed causing chaos, while another person said he set a tree on fire because he liked the smell of burning leaves.

Although rare, cases involving firefighter arson have received extensive media attention that is probably not reflective of the profile of an arsonist. In November 2024, for instance, a volunteer firefighter with the Medford (NY) Fire Department was arrested for allegedly setting a brush fire in a wooded area of Suffolk County. The suspect, a twenty-year-old man, was suspended from the department and faced termination if convicted. Although the fire burned a limited area, it came on the heels of nearly a dozen wildfires in New York, including the Jennings Creek Fire on the border between Passaic County, New Jersey, and Orange County, New York. A few months earlier, a 38-year-old off-duty firefighter with the California Department of Forestry and Fire Protection was suspected of starting five separate fires totaling about 1 acre of wildland in Sonoma County. Another young firefighter was arrested in 2023 in Yavapai County, Arizona, suspected of setting eight fires, including four wildfires, in two neighboring counties. And a 44-year-old volunteer fireman faced twenty counts of felony arson for multiple brush fires set in Alpine, Texas, in 2023, identified through DNA evidence and cellphone data.

Debris and Open Burning

In many parts of the country, especially in rural areas that do not have an adequate waste management system, individuals use burn piles or burn barrels to get rid of household trash or yard debris, often without an appropriate permit or regard for weather conditions or suitable tools to prevent runaway or escaped fires. The Wisconsin Department of Natural Resources has noted that debris burning is the number one cause of wildfires within the state, usually during the spring when the snow has melted and people do yard cleanup. The debris often consists of leaves that have accumulated on the ground, pine needles, and brush. In Texas, a study by the Texas A&M Forest Service determined that 75 percent of the top preventable human-caused wildfires were due to debris burning, most of them during hunting season from September through January.

Equipment and Vehicle Use

Another type of ignition relates to motor-driven or mechanical sources, both small (welding equipment or chainsaws, for instance) and large (bulldozers, passenger cars, and recreational vehicles). The source of the wildfire may be metal/pavement contact such as a tractor dragging chains, vehicles parked along the roadside in a grassy area because of mechanical failure or sightseeing, collisions, or faulty electrical systems. In Oregon, cars were the number one cause of wildfires during the summer of 2021. These types of wildfires can also be ignited by somewhat unusual, accidental incidents. The 2024 Airport Fire in Southern California's Trabuco Canyon, for example, was caused by a crew from the Orange County Public Works Department that was, ironically, moving boulders so that off-road vehicles were prevented from entering the dry brush area. A spark from the heavy equipment ignited the blaze, which burned 23,526 acres. The 2024 Borel Fire was a large and destructive wildfire south of Lake Isabella in Kern County, California, that burned a total of 59,288 acres before being contained. The fire was determined to have been caused by a fatal car crash in which the vehicle caught fire and spread to the surrounding hillside.

When the human source is identifiable, legal action can sometimes be taken, as was the case with the 2021 South Fire near the San Bernardino National Forest near Fontana, California. The federal government filed a civil suit against a contracting company in 2024 for negligence after the steel treads of one of the company's excavators contacted rocks and burned vegetation, causing a 680-acre fire that damaged or destroyed twenty-eight structures. Prosecutors in the case alleged that the contractor did not use reasonable care in the operation of the equipment. The lawsuit sought $2.2 million for reimbursement of fire suppression costs.

More common is the wildfire sparked by someone driving a vehicle with a flat tire or on a wheel rim or where the driver pulls off the road and a hot tailpipe or catalytic converter ignites tall, dry grasses. Sometimes, only a small amount of roadside vegetation burns, but that is not always the case. The 2018 Carr Fire began when a vehicle had a flat tire and the car's wheel rim scraped against the asphalt, creating sparks that ignited the fire. The blaze burned over 359 square miles and 229,651 acres over two California counties. Nearly 2,000 structures were destroyed or damaged, half of them homes, costing nearly $1.7 billion in insured losses and suppression costs. Eight people were killed.

Firearms and Explosives

Ammunition and fireworks are two common ignition sources which can lead to wildfires in areas with dry vegetation, especially in restricted areas or during periods of high fire danger due to gusting winds or drought. Some types of bullet fragments, for instance, can reach temperatures as high as 1,472 degrees Fahrenheit, igniting organic material. Flare guns, signal flares, and road flares can cause fires when mishandled or placed improperly. Fireworks, whether ground-based, handheld, aerial, or explosive, can cause both human injury and property damage. Even smaller ones such as sparklers can emit flames or cause sparks to land in flammable vegetation, while larger ones like mortars can explode. Some aerial fireworks create flaming particles that can be spread by the wind and land outside the immediate area from which they are launched. Local governments have attempted to reduce wildfire risks by restricting or prohibiting the use and possession of fireworks, but they remain popular during holiday periods in many communities.

The federal government prohibits fireworks on public lands managed by the Forest Service, National Park Service, and Bureau of Land Management but those regulations do not stop individuals from using them in a reckless manner. One study cited in *High Country News* found that nearly twice as many wildfires were recorded on July 4 as almost any other day in the West, as many as 18,500 of them. On July 4, 2024, fireworks ignited the Tea Time Hill Fire in the Wharton State Forest in southern New Jersey, burning 4,300 acres. That same day, the Hawarden Fire, started by illegal fireworks, burned only 600 acres in Riverside, California, but destroyed six homes and damaged eighteen others, causing $28 million in damage and $1.5 million in response costs over eight days. Three sixteen-year-old teenagers were captured on video fleeing after fireworks ignited the fire.

Misuse of Fire by a Minor

There is a special category of fires caused by a person seventeen years of age or younger. The old excuse of a child "playing with matches" out of curiosity is still used today, but the type of device varies. Sometimes a juvenile will seek out a location where there is unlikely to be any adult supervision or attention, or a spot like a playing field where there is easy access. Older minors, especially teenagers, may be engaged in acting out behaviors, or have a destructive nature or want to start a fire to express anger toward someone or something. There are also cases when juveniles mimic an adult's behavior like repeatedly flicking a lighter but with destructive results.

Even juveniles face penalties, although it is less likely that they will pay monetarily in cases of damage. A thirteen-year-old girl started the 2014 Cocos Fire that destroyed forty buildings, causing $5.7 million in damage in San Marcos, California, resulting in a trial where the judge concluded she had maliciously and intentionally started a fire in her backyard. Convicted of arson, she was ordered to pay $40,000 in restitution, sentenced to 400 hours of community service, and told to write letters to all of the fire victims. The case was not unusual. A 2018 fire in Oregon's Columbia Gorge burned almost 47,000 acres after a fifteen-year-old boy threw fireworks into a canyon. He wrote an apology to everyone living in the gorge, was sentenced to 1,920 hours of community service, and five years' probation. The court also ordered him to pay $36.6 million in restitution under the state's delinquency statute, which the circuit court judge said was clearly proportionate to the offense because it does not exceed the financial damages caused by the youth. Under the sentencing, the judge authorized the Hood River County Juvenile Department to create a payment schedule for the exceptionally large amount. After ten years, if the boy successfully completes probation, does not commit an additional offense, and complies with the payment plan, the court can grant full or partial satisfaction of the restitution judgment.

In an unusual twist of punishment, the State of California sought $675,178 from the parents of two children who set the 100-acre Skyline 3 Fire in San Diego County in June 2020 during a family camping trip. In a lawsuit from the state's Attorney General filed three years after the fire, the adults were accused of failing to supervise their children who started the fire, as well as another fire three days earlier. The fire caused mass evacuations in the area. The amount of the lawsuit represented the damage done and the cost of fighting the fire.

All fifty states have some sort of law that holds the parents or a child's legal guardian responsible for the damage or injury that results from their children's intentional or willful actions, including arson. There are, however, various conditions or limits attached to parental responsibility, including a dollar ceiling and whether or not homeowners' insurance can be applied. In some states, parents could be charged depending on whether or not fireworks were the cause of a wildfire, or the type of land, such as a forest, where the wildfire occurred. Some cases will be heard in juvenile court, while others are dealt with through mediation. Generally, the goal is to encourage parental supervision of children by making adults financially responsible for what their children do wrong.

Power Generation, Transmission, and Distribution

In recent years, there has been increased attention to wildfires caused by electrical power sources, even when the source is wildlife, such as a bird or squirrel that comes into contact with high-voltage equipment. The deadly wildfire on the Hawaiian island of Maui on August 8, 2023, exemplifies the damage done by a combination of high winds caused by a hurricane and periods of drought, combined with broken utility lines. The fire burned nearly 6,700 acres on the island, including 2,170 acres that devastated the town of Lahaina, where most of the 102 deaths occurred. In October 2024, a report on the investigation by the Maui Fire Department and the Bureau of Alcohol, Tobacco, Firearms and Explosives on the origin and cause of the fire revealed that the re-energization of the lines caused sparks to fall to the base of a utility pole, igniting vegetation below, along with arcing of overhead lines between two poles that resulted in a line falling to the ground, further igniting vegetation. Although classified as accidental in the report, seven defendants were named in 450 subsequent lawsuits brought by more than 2,200 claimants.

In an unprecedented legal settlement just a year after the fire, a $4 billion compromise agreement was reached that protected the defendants in the case from future litigation but also allowed future plaintiffs to apply for access to settlement funds. Nearly $2 billion of the money will come from Hawaiian Electric, which serves almost all of Hawaii's 1.4 million residents. The utility company was uninsured for a fire of this scale and was criticized for failing to shut down its power lines during the dangerous wind conditions associated with the fire. Another defendant, Kamehameha Schools, was criticized for failing to maintain the vegetation in the area where the fire started.

The Maui fire and the accompanying agreement far surpassed the $80 million that utility firm Southern California Edison (SCE) settled with the US Forest Service in 2024 over the 2017 Thomas Fire. The 280,000-acre blaze killed two people in Santa Barbara and Ventura counties in Southern California, destroying more than 1,000 structures and damaging forest land in the Los Padres National Forest. The company had said that its equipment might have sparked the fire, but did not admit any wrongdoing. It was just one of several suits in which California utility equipment has been alleged to have started wildfires, including three in 2019 in which SCE agreed to pay $360 million to settle claims with cities and counties. SCE equipment ignited nearly 60 percent more fires in 2024 than in 2023, although most were very small and quickly extinguished. The California Public Utilities Commission requires utilities to report all ignitions when their equipment is involved and the fire spreads more than one meter.

California utilities have been exceptionally hard hit by wildfires blamed on their equipment or operations. One of the earliest fires known to have been caused by utility equipment took place in September 1970. The Laguna Fire in San Diego County burned 175,424 acres and killed eight people after an oak tree hit power lines that came down during gusts from Santa Ana winds. The fire caused $234 million in damages (not adjusted for inflation). One of the important legal cases dealing with utility company liability involved San Diego Gas & Electric (SDG & E) and three blazes that led to the evacuations of a half million people, over 200,000 acres burned, and at least 1,500 homes destroyed, with two persons killed. In a ruling by the California Public Utilities Commission (CPUC), the utility was held responsible for the October 2007 Guejito, Witch, and Rice fires by failing to properly clear vegetation near its equipment as required by state law. Although the utility did not admit liability, the company paid over $2.4 billion in settlements in 2,500 lawsuits filed for damages, and $14.3 million to the state. One set of property owners sought $130 million for the 23,000 trees that were destroyed, including thousands of endangered Englemann oaks native to southern California. The legal battle did not end there. Ten years after the fire, administrative law judges ruled that the utility did not reasonably manage its facilities, and that the wildfires were not outside SDG & E's control. As a result, they ruled the company could not pass $379 million in costs that were not covered by its own insurance along to its customers.

Pacific Gas and Electric (PG & E), California's largest utility, has borne the most intensive scrutiny and faced the most financial penalties in California's wildfire liability cases. In 1994, the company was found guilty in 739 counts of negligence for not trimming trees near its power lines that led to the Sierra Fire in the state's gold country. The company faced a $125 million fine by the CPUC in 2021 for the company's role in the 2019 Kincade Fire in Sonoma County; and $45 million in penalties after a tree fell and hit the company's equipment in 2021's Dixie Fire, in which nearly a million acres were burned across five Northern California counties, destroying more than 1,300 structures. PG & E also pleaded guilty to counts that included involuntary manslaughter in the deaths of eighty-five persons for unlawfully starting the 2018 Camp Fire in Butte County, California. An estimated 18,000 structures burned due to weather conditions, PG & E's equipment failures, and downed power lines. A $522 million settlement was eventually reached, including $270 million to the town of Paradise and $252 million to Butte County.

Multiple lawsuits, including one filed by Los Angeles County, blamed SCE for the deadly Eaton Fire in Altadena, California, in 2025 that burned 14,000 acres, 9,414 structures, and killed seventeen persons. Residents living in the foothills of Eaton Canyon, and a nearby security camera at a gas station, recorded videos of high-voltage transmission towers on the hillside with flames at the base. Two of the lawsuits that were initially filed claimed that SCE was negligent in maintaining its equipment, while a class action suit alleged that an improperly grounded wire sparked fire in the surrounding vegetation that the company had failed to maintain. State and county fire inspectors did not find Edison responsible for the blaze, but other entities also filed suit against the company. The utility company acknowledged that videos have suggested a possible link between the utility's equipment and the fire, noting that there had thus far been no evidence of broken conductors, arch marks, or evidence of faults on energized lines in the area where the Eaton Fire had started.

Among the major wildfires in other states attributed at least in part to a utility is the 2011 Bastrop County Complex Fire near Austin, Texas, that destroyed 1,700 structures, leading to $350 million in insured losses. A class action lawsuit was filed against a private tree company and Bluebonnet Electric Cooperative, among other entities, alleging that the firm had failed to prune trees that damaged power lines, sparking fires during known risks due to weather conditions. It took nine years for the parties to agree to a $5 million settlement. That figure is in stark contrast to the $178 million utility company PacifiCorp agreed to pay in June 2024 to resolve claims by 403 plaintiffs from the Labor Day 2020 Fires in southern Oregon and northern California, and the Beachie Creek and Echo Mountain fires in northwestern Oregon.

Railroad Operations and Maintenance

Railroad track maintenance, including clearing vegetation and rail grinding, was once a major cause of wildfires in the United States. A 1923 *Journal of Forestry* article noted that "it is common knowledge that in the United States steam railroads are one of the principal causes of forest fires" (Hawley, 1923). The author points out that this is especially true along the Atlantic Coast due to the combination of wooded areas with intensive railroad development. One hundred years later, in August 2023, rail grinding once again led to dozens of small fires in Lancaster, Pennsylvania, along a rail line owned by the Norfolk Southern Port Road rail line. Track grinding is typically done to remove irregularities from worn tracks to remove warping due to use and friction. Railroads use water cannons that soak the wood ties in front and back of the grinding machines to put out fires caused by sparks from the grinders, but in this instance, that equipment was too far down the track to deploy quickly enough.

While less likely today, railroad fires still take place, as in June 2024 when sparks from equipment used to cut a section of rail near Wickenburg, Arizona, started a wildfire that destroyed fifteen structures at a cost of nearly $1 million to fight over the course of five days. Investigators and officials from the Arizona Department of Forestry made the determination of the cause of the fire because of how the blaze moved upslope away from the tracks.

Recreation and Ceremony

An improperly attended or extinguished campfire is a common cause of wildfires, but there are many other types of activities that fall into this category. Bonfires, ceremonial fires, barbeque fires, and fire pits that are not built or placed properly can inadvertently lead to out-of-control wildfires. Certain types of lanterns, such as luminaria, may be knocked over in the wind if placed on the ground, or may land in dry vegetation if launched aloft. These miniature hot air balloons are made of paper, plastic, or lightweight wood, and sometimes travel miles from the launch location, reaching several thousand feet into the air. Even when an attempt is made to extinguish a recreational fire, the remaining coals or ash may not be discarded in the right type of receptacle or may smolder for an extended period without attention.

Increasingly popular gender reveal parties, which often involve a colored smoke bomb, are also a potential ignition source. In April 2017, at a party in Arizona's Coronado National Forest designed to tell guests that a baby was going to be a boy, the baby's father fired his gun four times before hitting an explosive device that was filled with blue dye. The resulting explosion set the grass ablaze, becoming the Sawmill Fire that burned 46,991 acres over an eleven-day period and cost over $8.2 million in suppression costs. The baby's father, who pleaded guilty, was sentenced to five years' probation, and eventually agreed to pay $220,000 in restitution, along with agreeing to appear in a public service announcement regarding the fire.

Another human cause of wildfires occurs during hunting season, when individuals hoping to ward off the outdoor chill start fires to keep warm. This also can happen when there has been a dry season during the summer and fall when grasses and brush become dry and flammable, and hunters fail to put out warming fires. In November 2024, a fire started in thick, dry marsh grasses near a pond in an inaccessible area near California's Salton Sea. Officials said the fire, which burned 400 acres of wetlands, was likely due to duck hunters in the popular area. The fire also caused damage to wildlife, including waterfowl, rare burrowing owls, and pupfish.

Smoking

Discarded cigarettes, cigars, matches, and similar combustibles are a common source of wildfires, although the circumstances vary. Usually, it is not the butt itself that ignites, but rather the embers from tobacco, especially when dry leaves or grasses are present. It has been theorized that when automakers stopped putting ashtrays in cars, smokers began casually tossing cigarettes out of their car's window. Even snuffing out a cigar or cigarette on the pavement, a log, or other surface can start a blaze. Public service announcements against littering and Smokey Bear posters and ads were one attempt to stop the practice, although it still continues. In 2022 near Ellensburg, Washington, a witness saw a car speeding down the road with a cigarette butt protruding out of the car's window. Later, a fire investigator walking along a stretch of the highway discovered a clear line in the brush indicating where the fire had started, igniting the Vantage Highway Fire that burned more than 30,000 acres, taking eleven days and costing over $7.5 million to extinguish.

Other Human Causes of Wildfires

The NWCG includes several types of sources of wildfires in this final category, ranging from electric fences to spontaneous combustion. Structure fires can also become wildfires when the fire begins with a chimney, or during a power failure. This category might include examples like the 2018 Ranch Fire in northern California, started by a man who used a hammer to drive a metal stake in the ground when installing a shade cloth on his property, causing sparks. The resulting wildfire spread over four counties. More than 410,000 acres burned, with one firefighter killed. One of the state's deadliest fires, which killed fifteen people, was California's 2003 Cedar Fire, caused by a hunter who decided to start a signal fire to get help. That fire burned 280,278 acres and destroyed 2,820 buildings.

Recent studies have examined the danger posed by oil and gas wells in wildfire burn areas. There are an estimated 100,000 wells in the nineteen states west of the Mississippi River that have burned in recent decades and are likely to do so in the future. The majority of them, about 87 percent, are in California, and nearly 3 million Americans live within 3,200 feet of the wells. The problem is especially acute in Orange, Kern, Fresno, and Los Angeles counties, places where the number of people and wells are both high. Texas, Oklahoma, and New Mexico have wells in high fire risk areas, and the problem is expected to get worse as wildfire activity and population both expand. By 2050, more than 122,000 oil and gas wells are expected to be in high wildfire risk areas, and by late century, that number will be more than 205,000. In addition, people of color face a disproportionate risk of exposure to wells in the path of wildfires, and new wells are disproportionately being drilled in places where Black and Latino people live. Even when wildfire is absent, leaks of flammable gases pose a health risk.

In some cases, the causes of wildfires may be difficult to determine, especially if there is more than a single origin point. In the December 30, 2021, Marshall Fire in Colorado, for instance, nearly 300 lawsuits were filed as the cause of the fire was under dispute. The wildfire killed two people, destroyed a thousand homes, and caused more than $2 billion in property damage. Was the fire caused by members of an organization who burned trash and yard debris in a pit on their private property, covering the pit with about a foot of dirt that later blew away in hurricane force winds? Or did the fire stem from a power line that came loose during those same winds, causing electrical arcing, with hot pieces of metal flying into dry grass? Or were two telecommunications companies that had their equipment on the same pole as the utility company to blame? County officials believed an unmoored power line from Xcel was one source of the fire, but that claim was disputed by the utility company. Fire officials eventually determined that the fire was human-caused, with several potential starting points identified.

Natural Causes of Wildfires

There are two natural causes of wildfires: lightning and volcanoes. Historically, lightning was counted as a major cause of wildfires, but that appears to have changed. An analysis of wildfires on US Forest Service lands found that lightning was responsible for just 5 percent of California's wildfires, compared to 95 percent that were human-caused. Those caused by lightning were due to "hot" lightning that generates intense heat, sometimes referred to as electrical fire. The damage lightning does to the natural environment, such as the trees in the forest, varies by species. The pines of Arkansas, for instance, may have a 70 percent mortality rate when struck by lightning, whereas the Ponderosa pines of the Kaibab Plateau in northern Arizona may face destruction ranging from 1 to 33 percent. Even cacti, such as southern Arizona's saguaros, can be electrocuted, as can the cabbage palms of Florida, vegetation such as grasses, and even crops. Because lightning fires occur randomly, from a geographic standpoint, in comparison to human-caused fires that occur in accessible areas, the cost of fighting them is much higher.

The strikes are usually influenced by weather conditions, especially when temperatures are very high and precipitation and humidity are low. Strong dry winds also can lead to cloud-to-ground lightning. The NIFC reports that between 2001 and 2023, lightning-caused wildfires were highest in number between 2001 and 2007, with 16,165 in 2006, decreasing after that to less than 10,000 since 2012. The fewest wildfires due to lightning were generally in Alaska, Southern California, and the eastern region of the United States, with most occurring in the Northwest, Northern Rockies, and Rocky Mountains. The Dragon Bravo Fire, which started July on 4, 2025, at the North Rim of the Grand Canyon in Arizona, is a notable example of the damage that can happen when drought and low humidity, high winds, and high temperatures occur. The fire started as a lightning strike, quickly engulfing the historic Grand Canyon Lodge and more than 70 other structures. It became a megafire (over 100,000 acres) in August 2025 as it continued to burn. Initially managed as a "confine and contain" fire as officials allowed it to burn naturally under supervision, high winds shifted strategy to full suppression as the fire surged to become one of the state's largest fires.

At first glance, volcanoes might not seem like a likely cause of wildfires; most people associate volcanic activity with Hawaii due to ongoing eruptions from Hawaii's Mount Kilauea. But the 169 potentially active volcanoes in the United States identified by the US Geological Survey are in eleven states, with the majority in Alaska (see Table 1.1).

Table 1.1 Number of Potentially Active Volcanoes in Each State

Rank	State	Number
1	Alaska	141
2	California	18
3	Oregon	17
4	Washington	7
5	Hawaii	5
6	Utah	4
7	Idaho	4
8	New Mexico	3
9	Arizona	2
10	Nevada	2
11	Colorado	1
12	Wyoming	1

Source: Data from US Geological Survey. https://www.usgs.gov/vhp.

Fires from volcanic activity (FFVA) are a seldom studied phenomenon, despite the fact that they increase both the area and the severity of impacts beyond the initiating volcanic activity. Lava flows have a temperature range of from 700 to 1,200 degrees centigrade, and environmental factors, such as high winds, fuel availability (such as vegetation), and drought or prolonged dry conditions can increase the likelihood of fire ignition and spread. During the Maunaulu eruption from May 24, 1969, to July 22, 1974, fires from the lava flow encroached on ohia lehua forest that is believed to be from 200 to 750 years old. The deadliest, or some might term worst, damage from a volcano began on April 30, 2018, when the Puʻu ʻŌʻō cone of Kīlauea collapsed, triggering an eruption that would last for four months. By the end of the eruption, 716 homes were destroyed, and some 3,000 residents were displaced. The lava caused an estimated $236.5 million in damages to public infrastructure, $27.9 million in farm losses, and $415 million in tourism revenue losses.

Approaches to Wildfire Management

Most historians have approached wildfires with narrative timelines identifying the evolution of human history and our approaches to coping with fire. As explained earlier in this chapter, the historical record begins with indigenous peoples' practices related to fire, but the public's interest in wildfires, and policies related to managing fire come later in US history. During the nineteenth century, when large fires threatened property and resources, the public began to realize the real dangers that wildfires posed both economically and in terms of human health. There were two distinct fire periods in Maine in August–September and October 1825, for instance, that caused an estimated $10 million in damage to state forest lands but no persons were killed. That fire (and one that burned 3.8 million acres the same day in the neighboring Canadian colony of New Brunswick, killing 160 persons) captured the public's attention and is considered the beginning of the modern era of wildfires.

The focus during most of this period was on wildfires in the Northeast because the United States began to experience (and record) large wildfires that differed in scale to the ones used by indigenous people. The October 8, 1871, Peshtigo Fire that covered over a million acres killed over 1,400 people in northeastern Wisconsin, and Minnesota's Hinckley Fire killed 418 more. The fires were characterized as forest fires because they led to the loss of valuable timber. Sometime around the 1890s, the federal government's General Land Office, established in 1812 to dispose of public lands, hired rangers who worked during the fire season with responsibility for notifying local residents if a fire headed their way. But they did not have responsibility for fighting fires at this time, and the fires generally were allowed to just burn.

Fire management could be construed as a side effect of the Progressive era and the conservation ethic of planned and efficient progress. By the turn of the century, the practice of disposing of public lands into private ownership was evolving into the creation of areas set aside for preservation and protection. With the designation of forest preserves, national forests, national parks, and wilderness areas, the responsibility for protecting those areas fell to the states.

The fires of 1903 and 1908 burned an area from New York to Maine estimated to be more than a million acres, much of it in the Adirondacks. Lacking a firefighting regime at the time, workers were imported to fight the blazes, some of which were intentionally set. This series of events, among others, were sufficient to move state and local governments on a path toward fire suppression. Some involved partnerships with timber companies, such as the lookout system set up in Maine and Vermont. Some timber companies had their own firefighters; some localities had fire wardens or fire officers. Fire codes or restrictions were seldom the norm, and where they did exist, they were at best lenient.

The national response to fire management could be said to have started in 1876 when Franklin Hough was hired as a Special Agent in the Department of Agriculture, tasked with assessing the nation's forests. When the Forest Reserve Act was enacted in 1891 that allowed the president to set aside public lands and to exclude them from settlement, protecting the timber on those lands was not yet a major concern. With the Transfer Act of 1905, Congress took the responsibility for protecting the forest reserves away from the General Land Office within the Department of the Interior, giving those duties to the Bureau of Forestry in the Department of Agriculture—the place where crops (like timber) are managed. The newly created US Forest Service, under its first chief, Gifford Pinchot, inherited fire control as part of its working portfolio, with the period from about 1910 to 1930 the time when the new agency tried to make its way toward becoming a fully functioning bureaucracy.

Wildfire Suppression and Exclusion

Public concerns about wildfire damage and deaths shifted to the West in 1902, when Washington's Yacolt Fire burned more than a million acres and killed thirty-eight people. In 1905, the US Forest Service hired rangers to fight fire and created professional firefighting standards. They also built platforms above trees, structures that later became lookout towers. But if there was a singular year that changed the country's approach to wildfire, it was 1910, the year of "The Big Blowup" that author Timothy Egan (2009) called "The Big Burn." Fires in Montana and Idaho killed over eighty firefighters and torched more than 3 million acres at a cost of over $1 million. The fires were partially responsible for the passage of the Weeks Act (excerpted in Chapter 5) in March 1911 that authorized firefighting matching funds for state forest fire protection as well as the federal purchase of more acreage to add to the National Forest System.

From about the turn of the century until the 1930s, wildland fires were managed on an "as-needed basis" as the USFS focused on developing a bureaucratic structure and organizational policies. California's systematic fire protection became the model and Coert DuBois, profiled in Chapter 4, provided the handbook in his 1914 *Systematic Fire Protection in the California Forests.* Under the 1924 Clark-McNary Act, states became working partners with the USFS after the agency conducted an inventory of the states' fire protection needs. By 1949, the country's needs were estimated at $20 million, and the federal government, under Clark-McNary, was willing to pay for half that amount.

More conferences and experimentation with various management practices ensued until 1935 when the Forest Service's 10 a.m. policy was adopted, becoming a major tool of suppressing and then excluding wildfires. Under the new policy, once reported, fires were to be contained and controlled by 10 a.m. If that goal could not be met, the goal was shifted until 10 a.m. on the following day. The policy increased the importance of an initial attack by firefighters using whatever equipment and tactics were available. During the Great Depression and until it was disbanded in 1942, the Civilian Conservation Corps (CCC), a federal work relief program, sometimes employed men as firefighters, primarily in the West. Enrollees, as they were called, served in a variety of positions, based to some extent on their work history if they had one. While some served as members of permanent fire patrols, most men had no firefighting background or experience. They did, however, become an easily mobilized force that could be used to fight fires. They also constructed fire lookout towers, trails, and roads that enabled men and their supplies to reach wildfires more rapidly. The men removed flammable material and constructed fire breaks. In New England, the CCC constructed water basins and ponds to give firefighters a supply of water for tanks and pumps. By the time the organization disbanded, forty-seven of the enrollees had died fighting fires.

The Second World War changed the direction of the USFS policies in several ways. The manpower that had emerged from the federal Emergency Conservation Work program was reduced by the need for men to fight the war. Civil defense programs and the military became affiliated with forest protection goals. Fears of mass fire shifted coverage to the urban fringe and conflagration control. After the war, the agency looked to military surplus for its equipment needs. It added rural fire protection to its portfolio in 1954. In 1971, its suppression policy was changed to a goal of limiting wildfires to 10 acres as long as they did not destroy or threaten private property or property adjacent to wilderness areas, becoming a modified "let burn" policy by the 1980s.

In the 1970s, the USFS found itself contending with a wave of environmental legislation, like the 1964 Wilderness Act and the 1970 National Environmental Policy Act, that put some limitations on its authority. This period was also a time when interagency rivalries over resources pitted the expectation of autonomy against the reality of partnership expectations. The National Wildfire Coordinating Group (NWCG) was one result as the Secretaries of Agriculture and Interior explored how to respond to the near breakdown of a coordinated response to the seventeen major fires that burned in Southern California over just thirteen days in 1970. The result was called FIRESCOPE—a system that collected and reviewed information about a fire to integrate resources and enhance communication. Less an acronym and more a model, it called on cooperation at every level of government.

Prescribed Burning

Considered a tool in the wildfire management toolbox, the terms "controlled fires" and "prescribed burns" are often used interchangeably. They involve setting a fire to maintain the health of a forest by using a plan or prescription of the size of the fire, what is expected to burn, and what will be accomplished. Typically, the managing agency will map where the fire is to be ignited, the management of smoke from the fire, how the public will be informed, what kinds and how much equipment will be needed, and standby resources. This plan may include broadcast burning across a swath of land or pile burning of stacks of vegetation like leaves, brush, and limbs, or slash leftover from logging or forest thinning.

The strategy has several advantages. Prescribed burns can reduce flammable materials that might otherwise result in a wildfire and have the additional benefit of helping certain types of species such as pine release their seeds. Used properly, these fires can reduce insect populations such as ticks, return nutrients to the soil, and open up the forest canopy to provide more sunlight and space for young trees and other plants. This type of fire has other uses besides ones that benefit forests. In April 2024, for instance, the US Army conducted a prescribed burn on its Schofield Barracks training range complex in Hawaii. After receiving approval and preliminary preparations with allied agencies, the Army's Wildland Firefighters burned highly flammable guinea grass and other vegetation where a wildfire could threaten more than thirty species of endangered plants and animals. However, there are certain risks posed by prescribed burning when sufficient planning has not taken place, or when conditions such as weather change dramatically without notice. In these cases, property can be damaged or destroyed, people and animals can be killed, and there may be unintended consequences such as traffic congestion, smoke that reduces air quality, damage to infrastructure, or liability concerns.

In recent years, prescribed burning has often been stymied by funding issues, evidenced by the Trestle Project in El Dorado County, California. The Forest Service had anticipated the dangerous fuel conditions in the area in 2000, proposing a 15,000-acre project of prescribed burning that was expected to be completed by 2020. The Trestle Project faced staffing shortages, pushback from environmental groups, and a lack of funding that resulted in only 14 percent of the acreage treated by August 14, 2021. That was the day that the Caldor Fire started as a small brushfire, but the USFS inexplicably pulled crews and engines off the fire in the early hours. Two days later, in a span of less than fifteen minutes, the fire leveled the community of Grizzly Flats but spared a nearby community, Kirkwood, where an escaped prescribed burn had taken place in 2019, according to the Center for Economic and Policy Research.

The acceptance of prescribed burning was further complicated in October 2022 when a USFS employee was arrested after an ember from a prescribed burn conducted on the Malheur National Forest in eastern Oregon landed on nearby private ranchlands. The crews quickly extinguished the small blaze but the local sheriff arrested the firefighter in charge of the burn on misdemeanor charges of "reckless burning." The prescribed burn incident took place in an area where armed protestors had previously taken over the Malheur National Wildlife Refuge in a nearby county known for being the setting for disputes over federal authority.

Private attorneys paid for by the US Department of Justice represented the employee, who was indicted by a county grand jury in February 2024, and the case was sent forward to federal court. The executive director of the Oregon-based Firefighters United for Safety, Ethics, and Ecology called the indictment "an extreme overreach of law enforcement" and the regional forester told *Inside Climate News* "I believe the courts will get this case right and will not single out a federal employee for carrying out his official duties. Our agency is committed to standing by [the employee] and all our team members who responsibly carry out their official duties in accordance within agency protocols" (Stringer, 2024).

The attorney representing the defendant argued in court documents that the employee could not be charged with a crime simply because someone disagreed with the USFS decision to conduct the prescribed burn, according to Oregon Public Broadcasting. The sheriff had told the grand jury that he speculated the agency had authorized the prescribed burn because it had been a slow year without much going on. "They were looking for something to do." The employee's arrest, the sheriff said, got "huge national exposure" (Wilson, 2024). In federal court, defendant's attorney asked for the case to be dismissed, and the district attorney who had initially presented the case to the grand jury did not oppose the motion, which was then approved by the judge in June 2024.

Attempts to return prescribed fire as a management tool resumed under the administration of USFS Chief Randy Moore in 2023. Under his leadership, the Forest Service's National Prescribed Fire Resource Mobilization Strategy outlined the agency's efforts to implement the Wildfire Crisis Strategy, which he called "a fundamental shift in how we do business" (US Forest Service, 2023). Chief Moore said the agency would move forward by prioritizing prescribed fire on a larger geographic scale, delaying or setting aside other mission-related work because the window for prescribed fire was limited, especially in the West. The status of Moore's plan to return prescribed fire to the federal management plan became uncertain, however, with the naming of Tom Schultz as the new Chief in the Trump administration. Unlike many of the agency's prior leaders, Schultz was not a career Forest Service employee. He previously served as vice president of resources and government affairs at Idaho Forest Group, where he led timber procurement operations.

One of the challenges the agency faces is a lack of qualified professionals capable of managing the types of specialized fires required for prescribed burns, referred to as "burn bosses" in firefighting circles. An estimated twenty states have responded to the need for specially trained burn managers by developing their own certification programs separate from training conducted by the USFS. These types of training programs, some of which have been in place for many years, vary considerably from one jurisdiction to another. Certification then allows practitioners to burn when others cannot, to burn longer through the day, with liability protection as long as the law's requirements are met.

Fire Prevention Campaigns

Most people are familiar with the name and image of Smokey Bear, the iconic animal, real and symbolized in posters and advertising. But are they aware that the United States embarked on efforts to warn people about the dangers of forest fires as early as 1902? That's when the General Land Office produced a poster telling the public how to keep forest fires from getting out of control. In 1939, an Uncle Sam-lookalike pointed at an inferno, warning "Your forest, your fault, your loss." Over time, the messages the public saw about forest fires changed only slightly, although the messenger would evolve as a result of the Second World War.

The first modern posters dealing with forest fires were produced in 1941, showing two men in overalls with shovels battling a blaze and the slogan "Forest Defense Is National Defense." The message in the drawing of the ominous fire was that putting out fires was necessary to provide the defense effort with a sufficient amount of timber. This was also the year that there were 208,000 forest fires in the United States that burned 30 million acres of land. Nine out of ten of those fires were caused by humans, and according to the USFS, they were also preventable. In spring 1942, just months after the bombing of Pearl Harbor, Hawaii, Japanese submarines fired shells that landed on an oil field near the Los Padres National Forest in California. The attack was the first on the US mainland, reminding Americans that the war was close to home, raising fears that such attacks could ignite massive wildfires that would threaten the nation's lumber supply.

Soon after, the USFS established the Cooperative Forest Fire Prevention Program (CFFP) in partnership with the War Advertising Council and the Association of State Foresters. The theme of the initial effort was definitely militaristic, rather than resource-based. The medium for the campaign was posters that resembled the wartime messages that were popular at the time. One message said, "Another Enemy to Conquer: Forest Fires," and a 1943 poster portrayed Germany's Adolf Hitler and Japan's Hideki Tojo peering down on a blaze said, "Our Carelessness: Their Secret Weapon" as a way of reminding the public of the importance of timber to the war effort.

The effort was enhanced by the August 13, 1942, American debut of the Walt Disney animated film, *Bambi*, in which a young fawn grows up in a sylvan setting after his mother is killed by a hunter. Bambi and his wildlife friends learn about the perils of the forest and flee a campfire that begins to spread, and at the end of the film, Bambi assumes the role as Great Prince of the Forest. Walt Disney allowed the popular characters in the film to be used for a one-year period, including a 1944 poster. Recognizing the value of an animal as part of the CFFP campaign, the group began looking for a replacement. The USFS turned to the advertising agency Foote, Cone, and Belding (now FCB), which remained on the wildfire prevention campaign.

Artist Albert Staehle is given the credit for his realistic drawings of a bear as the new mascot, named Smokey after a heroic New York City fire captain, Joseph "Smokey Joe" Martin. The character wore a campaign hat and jeans in its early incarnation, authorized on August 9, 1944. In 1946, Rudolph Wendelin became Smokey's official artist until he retired in 1973. In 1947 the bear was pictured with the slogan "Remember ... only YOU can prevent forest fires" and later, just the words, "Remember" or "Only you." The messages focused on the role of fire prevention, and according to Ellen Earnhardt Morrison (1976), whose history of Smokey Bear was endorsed by the USFS, the result of the advertising campaign was a "marked decrease in forest fires." An agency booklet, titled *Smokey's Record*, credited the program with saving America more than $10 billion in wildfire losses that did not occur (Morrison, 1976).

From an image to reality, in 1950 a group of Taos Pueblo firefighters rescued a black bear cub with burned paws from a fire in New Mexico's Lincoln National Forest, and the animal became the symbol of a new public relations campaign. The bear was treated for his burns and sent to the National Zoo in Washington, DC. Smokey's reputation and approval rating skyrocketed after the war. Singer Eddy Arnold recorded a song, "Smokey the Bear" in 1952, and commercial interest was sufficient that the government took the image out of the public domain and placed it under the Secretary of Agriculture. In 1964, the US Postal Service gave the bear his own zip code—20252—so that children could write him letters. In

1971, a television ad showed a picture of a Smokey Bear wristwatch with the voiceover reminding viewers, "By the time the second hand reaches 12, forest fires will have burned down another 100 trees unless you help." Smokey the cub died in 1976, and the bear's body was buried at Smokey Bear Historical Park in Capitan, New Mexico. Coverage included obituaries in numerous publications.

Contemporary Wildfire Messaging and Branding

The CFFP's message changed in 2001 to the broader tagline "Only *you* can prevent wildfires" reflecting changes in scientific knowledge about the beneficial effects of fire and the control of unwanted fire. Also changing was Americans' desire to have a piece of Smokey of their own, so in 2004, the CFFP produced the first set of guidelines on the use of the Smokey image and products. That same year, FCB brought back Bambi's support with public service announcements noting "Don't let our forests become once upon a time."

The popularity of a Junior Ranger campaign that began in 1952 did not always resonate with a younger cell phone-toting generation, and by Smokey's 75th birthday in 2019, his image had morphed into an animated emoji voiced by late night TV host Stephen Colbert and comedian Jeff Foxworthy. The government's (and Smokey's) message about wildfires also has tried to keep up with changes in technology. In 2008 Smokey entered the digital age with a Facebook page and social media accounts, a blog, and a mobile app. In 2012 a plush toy version of Smokey traveled aboard the Soyuz spacecraft to the International Space Station, according to NASA. Research by the Ad Council in 2018 showed that Smokey's image was recognized by eight out of ten Americans.

Over the years, there were calls for messaging that reflected the growing debate about fire prevention and forest policy. In April 1973, John McGuire, Chief of the USFS, stated, "Prescribed fire is for professionals only! We know of no place in the Nation where we can turn our backs on fire and let it run." When Morrison's book was published (1976), she referenced the Smokey Bear "detractors" and ecologists who argued that forest fires should "*not be prevented!* They claim that forest fires occur naturally and are needed to keep nature in balance" (emphasis in the original). Researchers blamed "the Smokey effect" for the Forest Service's fire suppression policies, noting that decades of overzealous firefighting had allowed vegetation and forest growth to accumulate, with the practice leading to more intense fires. Ecologist Chad T. Hanson has suggested that what he calls the atmosphere of fear and gleeful exaggeration has led to new and dangerous policies that support the megafire narrative and pyrophobia. The demonization of fire, he says, leads to the rollback of environmental laws driven by logging interests and their political allies and misguided and regressive solutions to what is a perceived forest fire problem where one does not actually exist. Instead, he argues, the rhetoric detracts from the resources that should be focused on climate change and carbon-polluting industries. The societal paradigm of fear, he says, encouraged by a steady barrage of government-funded disinformation, ignores the natural events and cycles that have been occurring in forests for tens of millions of years (Hanson, 2021).

In recent years, the USFS appears to have tried totally different strategies to develop and send a message to the public. In late 2023, for instance, the agency, Ad Council, and National Association of State Foresters released Public Service Announcements (PSAs) in the "Smokey Is Within" series. The message, according to the Campaign Development Office, "is to inspire individuals to look within themselves, channel their inner Smokey and keep educated and empowered to help reduce wildfires in America" (Ad Council, 2023). With a tone that sounded on the edge of Zen, the groups noted that "the new campaign reminds audiences that when we practice fire safety, Smokey is within us all, encouraging individuals to learn how to prevent wildfires." As part of the campaign, Smokey was given a more lifelike look, with the ad agency FCB rebuilding his costume from scratch. Animatronics were used in the ads on television and other visual media to move the bear's mouth and eyes, to "give true emotions to America's favorite bear" (Ad Council, 2023).

Fighting Wildfires with Aircraft and Fire Retardant

Wildfire management has changed in response to technology and innovation, adding new equipment and methods. In 1919 the USFS contracted with the Army Air Service to perform aerial patrols of national forests. Firefighters who parachuted from aircraft to remote fire locations, called smoke jumpers, were first tested in 1939 on the Okanogan National Forest in Washington, and then used in 1940 on Idaho's Martin Creek Fire. After the Second World War, surplus military equipment, including air tankers, were used to augment resources. When aircraft were used in the 1920s to fight wildfires, the small planes were equipped with containers of water that were dropped to suppress the flames, although the results were minimal and inefficient. The water often evaporated before it got close enough to the fires to have an impact, so agencies began adding different chemicals to water to make it thicker. It was not until the 1950s that air tankers began being used, equipped with chemicals that rendered fuels non-flammable, with water serving as the medium to get the retardant to the ground. The early forms of fire retardant used hydrating clays such as bentonite and borate, and then water thickened with gum. The process was used primarily to create firebreaks, and the chemicals were effective until heavy rains washed them away.

In 1954, a collaborative government agency research project, Operation Firestop, was launched at Marine Corps Base Camp Pendleton in California to study new technologies, including the use of chemicals as fire retardants. California wildfires had burned 140,000 acres during the wildfire season, and officials used the Marine base's vast acreage in the one-year program. The project found that one chemical, ammonium phosphate, was the most effective, with other chemicals less viable as active ingredients. The US Forest Service began using a phosphate-based chemical in the 1960s. In 1970, University of Montana scientist Aylmer D. Blakely expanded on the Operation Firestop experiments and developed what he called the superiority factor method to determine which chemicals ranked highest in overall effectiveness.

The industry continued to improve more than just the long-term suppression capabilities of retardant, examining features such as corrosion, how much of the retardant hit the ground, whether safe for extended-term storage and use, and later, visibility when dropped to the ground. For instance, when iron oxide was added to the mixture, it permanently stained whatever it came into contact with, including homes and firefighters, making it unsuitable for environmentally sensitive use and in the wildland-urban interface. In the 1980s, studies showed that a blend of ammonium sulfate and ammonium phosphates was found to be equal in performance to other combinations of chemicals but at less cost. The Department of Agriculture, in approving the use of the substances used by the USFS, says that what is now used is 85 percent water, 10 percent fertilizer containing ammonium sulfate, and the remaining 5 percent includes chemicals such as the colorant, which is now a bright neon pink. While this is the basic formula for retardants now used in aerial fire suppression, critics of their use question the 5 percent that is not required to be disclosed for proprietary reasons, or what one watchdog group called the "chemical black box" of undisclosed ingredients.

Recent research continues to point to the health hazards caused by retardant. In a study published in October 2024 conducted by a team from the University of Southern California, researchers found that between 2009 and 2021, wildfire retardant application in the United States released more than 400 tons of ten toxic metals, with the fourteen commercial products investigated containing high levels of at least one metal. The ten are considered elements of concern and regulated by the federal government due to known toxicity. The researchers cautioned, however, that there is more study needed on the environmental fate of the metals released into the environment and the extent to which they contribute to human and environmental health risk (Schammel et al., 2024).

Questions continue to be raised over the use of these chemicals, including ammonia, which can be highly toxic and potentially deadly to fish, and in large quantities, toxic to humans. The concerns have gained traction as fire seasons have grown longer and more aerial fire retardant is being used. It is difficult to get an accurate picture of how much retardant various agencies use, since some reports blend use on both public and private lands and over different periods of time. For 2023, the USFS reported about 6.2 million gallons were used on USFS lands, primarily in Arizona and California (US Forest Service, n.d.). Other sources report that the Forest Service dropped almost 53 million gallons on federal, state, and private land in 2021 (Brassill, 2023). In comparison, CalFire reported aircraft dropped 194 million gallons of various types of retardant from 2006 to 2024. During the three weeks retardant was used on the Palisades Fire in January 2025, CalFire reported its aircraft had made more than 280 drops across 20 square miles (Margolis, 2025).

Various environmental organizations, including the non-profit group Forest Service Employees for Environmental Ethics (FSEEE), continue to question the use of aerial fire retardant and whether the government has followed federal environmental laws in assessing its safety. In a series of lawsuits against federal agencies, courts have ruled that the USFS violated the National Environmental Policy Act (NEPA) and the Endangered Species Act (ESA), requiring the agency to create new policies for fire retardant use. One of those policies dealt with the establishment of avoidance areas—ones where aerial fire retardant could only be used if human life or public safety were threatened. The USFS has developed requirements for monitoring and reporting in conjunction with the US Fish and Wildlife Service and the National Oceanic and Atmospheric Administration (NOAA). Another requirement, based on the violation of the Clean Water Act alleged in FSEEE's lawsuit and a 2023 court ruling, required the Forest Service to acquire a permit before dropping fire retardant into waterways, but that was later changed to include waterways under the definition of avoidance areas. Trade groups, community leaders, and members of Congress have sought to have the permit requirement eliminated or to exempt firefighting agencies from needing a permit, arguing that bureaucratic delays increase the risk of wildfire damage. A coalition composed of California lumber companies, veneer plants, and biomass facilities are fighting efforts to restrict retardant use, arguing that they have lost hundreds of thousands of forested acres to wildfires. One county supervisor who lost his home in California's Camp Fire that killed eighty-four people commented, "A massive forest fire burning down a huge amount of acreage is probably worse for the environment" (Wigglesworth, 2023).

Supporters of aerial application of retardants still contend that the strategy is an important tool in protecting communities, especially in more isolated areas. By using aerial suppression, the spread of a wildfire can be slowed down until other firefighting equipment and personnel can be deployed. From an optics perspective, the images of aircraft dropping retardant on a wildfire are very convincing. For those not on the ground to witness the deployment of materiel, the neon pink spray provides visual evidence that officials are attempting to aggressively fight the fire threatening their homes.

Improving Forest Health

During the 1980s, the old-growth forests of the Pacific Northwest were the center of national forest policies, but not because of wildfires. Instead, debates raged over timber companies' practice of clear cutting, sales of federal public lands for logging, various forms of fuel treatments, and the loss of habitat for endangered species like the northern spotted owl. The emotionally charged issue, framed as owls vs. jobs, became the Northwest Forest Plan or NFP, in 1993. After a stakeholder summit in Portland, Oregon, convened by President Bill Clinton, the NFP reduced the amount of logging in the Pacific Northwest by about one-quarter of what it had been in the 1980s. Lawsuits and appeals created paperwork delays in the plan's implementation, but those strategies also had the effect of reducing timber harvest.

Partly in response to the growth of increased environmental awareness by the public, an expanded research agenda within the USFS, and the social movement that accompanied environmental legislation enacted in the 1970s, the USFS began to shift to a new goal of ecosystem management. For officials, this was a different way of looking at forest health that altered the agency's day-to-day operations. Rather than looking at trees as a commodity and forest products as an industry, historian Gerald W. Williams says the federal government reviewed management ideas about the "new Forestry" and "new perspectives" that had been talked about for decades. Chief Jack Ward Thomas (1993–6) and Chief Mike Dombeck (1997–2001) referred to long-term sustainability and a collaborative stewardship agenda that focused on national forest protection rather than spelling out a specific wildfire policy (Williams, 2005).

George W. Bush, elected in 2000, promised to increase timber production and to reverse another Clinton-era strategy, the Sierra Nevada Framework, that covered eleven national forests in California. In 2002, the USFS subsequently announced it would revise the framework agreement because it would be impossible to protect the region's forests from catastrophic wildfires. Restrictions on harvesting old-growth trees were eliminated in order to triple the amount of logging allowed, with an advertising agency hired to sell the new plan under the slogan "Forests with a Future: A Campaign to Protect Against Catastrophic Wildfire." The campaign took advantage of the 2002–3 wildfire season to convince members of Congress to support the Healthy Forests Restoration Act (HFRA) that focused on hazardous fuels reduction projects on 20 million acres in USFS and BLM lands, supporting recommendations that had been made by the Western Governors Association. Competing versions of the measure in the House and Senate were driven by dueling scientists and researchers whose vision of the wildfire threat and how to manage it differed completely. Some said the science behind lands afflicted by drought and managed by fire suppression was unclear, while others called for large-scale restoration-based fuel treatments instead of just thinning trees.

The attempts at compromise were affected by the October 2003 wildfires that were destroying hundreds of thousands of acres in southern California. Ironically, the devastation occurred in urban communities far from forests or on private lands that would not be covered under the proposed legislation. In San Bernardino, California, for example, fires raced through canyons and washes filled with non-native eucalyptus trees, bamboo, oleander, and olive trees, destroying houses whose backyards were overgrown and filled with flammable debris. The media jumped on these facts, urging Congress to spend money on the WUI, with the *Los Angeles Times* writing, with a sense of ironic foresight, "If we're going to be dumb enough to continue building in high-fire-danger areas, brush needs to be cleared to prevent the spread of killer blazes" (Lopez, 2003). President Bush signed the HFRA, as amended, on December 2, 2003, with promises linked to funding for forest thinning and further appropriations to fulfill fuels reduction goals. It did not please everyone, but it would shape wildfire policy for many years to come.

Forest thinning became the new normal for the USFS, with a three-fold goal of reducing hazardous fuels, improving watershed conditions, and improving wildlife habitat. Efforts were also made to build cooperative partnerships rather than stoking the bitter antagonistic approach that had characterized the previous decades. The emphasis on thinning and prescribed burns, combined with an emphasis on partnering, continued with the Trump administration's March 1, 2025, executive order that renewed the goal of increasing timber harvests at the same time environmental regulations were being rolled back. The action blamed "onerous Federal policies" for the country's reliance on imported lumber that exported jobs and prosperity and compromised US self-reliance, calling for a reversal of policies to protect national and economic security. Although specific timelines for action were identified in the executive order, the actual implementation procedures were vague, such as reducing unnecessarily lengthy processes and suspending actions that imposed an undue burden on timber production (White House, 2025).

Wildfire Agencies and Organizations

There are myriad types of groups involved in wildfire policy, and one way of identifying them is by explaining what they do. Another way is by looking at the federal system of government, and the level at which they function. Still another typology is to divide them on the basis of those under the jurisdiction of a government agency and those considered non-governmental organizations (NGOs). Unfortunately, those arbitrary labels do not fit well in this complex policy and operational arena. This section looks at a sampling of many types of agencies and organizations, along with partnerships, active in managing wildfires, research and science, and post-fire activities throughout the United States.

Responsibility for Fighting Fires

One way of looking at the role organizations relate to wildfires is to examine their jurisdictions and how they relate to other agencies or groups. For instance, the federal government is responsible for responding to wildfires that begin on federal lands, such as those managed by the US Forest Service within the US Department of Agriculture. The USFS carries out wildfire management and response across the 193 million acres of the National Forest System, while the Department of the Interior (DOI) manages wildfire response for more than 400 million acres of national parks, wildlife refuges and preserves, other public lands, and Indian reservations. There are also distinct geographical differences in the role of the federal government. In the eleven contiguous western states, nearly half of all land area is owned by the federal government. When wildfires burn across the region, they disproportionately burn on federal land, giving the feds the prevailing responsibility for wildfire management in the West.

State governments are responsible for responding to wildfires that begin on non-federal lands (state, local, and private), except for lands protected by federal agencies under cooperative agreements. Local governments may also be involved in wildfire management, depending on the jurisdiction. For example, in Texas, local county judges are responsible for suppression of wildland fires, as is the case in Colorado and Wyoming, where the responsibility may lie with the local county sheriff.

But the jurisdictional lines often blur when decisions about management are made. Professor Donald F. Kettl examined the case of the 2011 Texas wildfires that burned more than 3 million acres in the Bastrop Fire near Austin. A member of the state's congressional delegation badgered the USFS because the federal government had not pre-positioned a giant aerial tanker in the state that could be used whenever needed. When the fire erupted, the plane, which belonged to a USFS contractor, had been shifted to fires in California. There was an expectation from some that federal resources like the tanker were to be treated as on-demand assets. A Forest Service representative noted that the agency had agreed to help fight the Bastrop Fire even though only 0.1 percent of the fire was on federal lands, deploying its teams to assist "because our friends in the Texas Forest Service asked us to help." In addition, federal help was needed because the governor had cut funding for the state's own fire agency. Kettl notes that the worsening drought in the West has also made fire seasons more intense, with the public used to watching television footage of giant planes dropping gallons of water on blazes below. "They see those tankers as a sign of government action," he notes, "Local residents increasingly expect a response that is federal and instant" (Kettl, 2015).

This view gained credence with the outbreak of the January 2025 fires in southern California. The initial fire in Pacific Palisades was within the jurisdiction of the Los Angeles City Fire Department, but the first fire truck that arrived at the scene was from the Los Angeles County Fire Department. While officials dealt with criticism over the lack of sufficient staffing and pre-deployment decisions, numerous firefighters and resources were brought in from other agencies to deal with the immediate need for suppression and safety. In addition, fire units from more than a dozen states responded with firefighters and equipment, as did Canada and Mexico, other states' National Guard members, and law enforcement officers.

Responsibility for Management and Planning

There is an alphabet soup of lower-level organizations at the federal level whose personnel are responsible for carrying out the administrative management of wildfire *policy*, rather than boots-on-the-ground firefighting. The number of personnel involved in these functions might seem staggering since there are so many agencies involved.

NMAC stands for the National Multi-Agency Coordinating Group, one of many designed to provide national-level strategic coordination of firefighting resources. A different type of coordination is the responsibility of the National Interagency Coordination Center (NICC), which then mobilizes the resources needed for wildland fires. This agency, based in Boise, Idaho, is also one of the best sources for information on regularly updated national wildfire statistics, such as the number of incidents to date, the amount of acreage burned, and the number of personnel assigned. Created in 1965 as the Boise Interagency Fire Center (BIFC), with the USFS, Bureau of Land Management, and National Weather Service as its members, the center was a way for the three organizations to pool their resources and coordinate their fire suppression and planning efforts. Two other members, the National Park Service and the Bureau of Indian Affairs, were added in the mid-1970s, with the US Fish and Wildlife Service added in 1979. Reflecting its national mission, the BIFC's name was changed in 1993, becoming the National Interagency Fire Center. Its membership now also includes the US Fire Administration (USFA) and the National Association of State Foresters (NASF). Its mission has also expanded to include several non-wildfire disaster needs related to earthquakes, terrorism, hurricanes, floods, and volcanic eruptions.

Another intergovernmental organization, the Wildland Fire Leadership Council (WFLC), was established in 2002 with the goal of supporting the implementation and coordination of federal fire management policy. It differs from some of the other groups in terms of membership by including not only federal officials, but also those at the state, tribal, county, and local levels. One vision statement for the group noted that its purpose was to "learn to live with wildland fire," representing a change from prior policies that referred to fire suppression. Over time, the council has expanded its emphasis areas to include challenges such as global climate change, diversity, and community resilience.

One of the newest (and shortest-lived) government groups was the Wildland Fire Mitigation and Management Commission, created in December 2021 under the Bipartisan Infrastructure Law during the Biden administration. The fifty-member commission represented all levels of government, as well as representatives from the private sector, with a singular purpose: "to address the wildfire system holistically, with a goal of creating communities and landscapes that are resilient to wildfire as a natural and integral part of the nation's future." The commission produced two reports to Congress, the first of which (February 2023) recommended strategies to improve aerial firefighting efforts and equipment needs. The second report (September 2023) was a comprehensive review with 148 recommendations urging Congress to act quickly on all of them.

Like other policy issues, wildfire has its own set of professional, interagency fire operations and even terminology. The National Wildfire Coordinating Group (NWCG) was established in 1976 as an operational group led by the Department of Agriculture and the Department of the Interior. Through a Memorandum of Understanding, the goal was to avoid wasteful duplication of efforts among the participating agencies and to develop policies, standards, and procedures that could be agreed upon. The participating agencies include ten primary members and two associate members. The key elements of wildland fire management are developed by committees and subordinate groups that establish standards such as wildland fire position standards, information technology capability requirements, and standards related to specific operations, such as Fire Unmanned Aircraft Systems.

Research and Science

In 1998, Congress established the Joint Fire Science Program (JFSP) with two goals: providing the funding needed to provide managers and policymakers with studies associated with managing wildland fire, and developing partnerships with both practitioners and policymakers with the latest information on topics of interest to others with similar responsibilities and programs while reducing redundant efforts. This focus on scientifically based information is implemented through the Fire Science Exchange Network that promotes the exchange of data through workshops, publications, field tours, discussion forums, and activities which promote interaction.

Some of the agencies involved in wildland fire management have very specific duties that are somewhat unique to their main mission. The National Weather Service, a part of the National Oceanic and Atmospheric Administration, for example, operates the Storm Prediction Center that issues Fire Weather Outlooks across the continental United States. This data helps other levels of government by determining whether pre-existing fuel conditions, along with forecast weather conditions, pose a significant threat of wildfires starting or their spread. The agency was already understaffed when it lost about 20 percent of its employees in early 2025 as part of the administration's government restructuring. A former NOAA administrator said the job cuts would most assuredly affect the availability, frequency, and accuracy of weather warnings.

In May 2025, the federal government retired a specialized database that analyzed weather and climate disasters with costs of a billion dollars or more that had taken fifteen years to create. The head researcher of the Billion Dollar Weather and Climate Disaster Online Database resigned, eliminating a valuable resource regularly used by insurers, scientists, and citizens, including those seeking a final tally of the cost of the LA wildfires.

At the state level, smaller agencies with very specialized portfolios have been created as problems are identified. In July 2021, California created the Office of Energy Infrastructure Safety (OEIS), previously known as the Wildfire Safety Division. The department is responsible for stopping the electric grid from causing wildfires by proactively conducting inspections of utility equipment. Originally located in the Public Utilities Commission in 2019, the move expanded the department's responsibility for tracking utility compliance with their own fire prevention plans, adding enforcement authority with the move to the Natural Resources Agency. The OEIS gained additional stature after the 2025 fires in California as policymakers examined the causes of those fires and the need for hardening the system. There is one more level of bureaucracy with the OEIS. A state law enacted in 2019 created the California Wildfire Safety Advisory Board, a panel of experts to advise the OEIS on utility company Wildfire Mitigation Plans (WMP), making recommendations to reduce the risk of utility-related wildfires.

Specialized Partnership Programs

Aside from the departments and agencies one might expect to find associated with wildland fire, there are others run by the federal government, or in partnership with non-governmental organizations that contribute to the effort to professionalize and improve the nation's firefighting capacity. One of these programs, Feds Hire Vets, run by the federal Office of Personnel Management, is designed to capitalize on the skills of military veterans that are applicable to fire and aviation management. Special hiring authorities allow agencies to take advantage of veterans' skills as they transition into civilian life.

A similar type of program is a joint effort between the California Conservation Corps, Conservation Legacy, and the Student Conservation Association partnering with both state and federal agencies—the Veterans Fire Corps. This retraining program provides a pathway for veterans to use their experience in the military to move on to positions within the fire protection field. This might involve gaining practical field experience in fuels reduction or prescribed firefighting, as well as associated jobs such as educational and public outreach.

Non-Governmental Organizations

While it might seem as if there is a sizeable bureaucracy of acronym agencies and departments within the government that deal with wildfires, there are even more non-governmental organizations with important roles. These groups, which are separate from any level of government, operate independently, and most are part of the non-profit sector, which means they are not subject to taxation. These groups are designated under the Internal Revenue Service code as 501c(3).

Unlike some other policy areas where there are often hundreds of non-governmental organizations associated with the topic or problem, such as hunger, animal welfare, and recreation, wildfire-specific groups in the non profit sector are less common. This is in part because of the specialized nature of the profession, and because the field itself is relatively new on the policy agenda. Some are professional membership organizations that stem from a broader base, such as the Society of American Foresters, and some include academicians engaged in fire research. One of these groups, the Fire Research Institute, was formed in 1983 as a repository for wildland fire research. Its database of materials is accessible online. A newer group, the International Association of Wildland Fire, was founded in 1990 as a professional membership organization that encompasses experts in all field of wildland fire management. It is also somewhat unique because its membership is not just national, but also global, and it serves as the editorial home for the *International Journal of Wildland Fire.*

Another sector of non-profits associated with wildfires consists of those groups whose mission is focused on post-fire restoration, such as replanting. Some, like Oregon Parks Forever, were established after a disastrous fire event, which in this case was the 2020 fires in the state. Other restoration groups have a national or even international focus, such as the National Forest Foundation and American Forests. Perhaps the most visible segment of non-profits, though, are the ones with a broad mission of which wildfire is only one part. For instance, the American Red Cross and World Central Kitchen provide services to persons displaced or needing assistance, part of a far-reaching sector of groups, large and small, that spring into action as needed. Best Friends Animal Society, based in Kanab, Utah, helped rescue hundreds of animals abandoned or lost after the 2025 California wildfires in partnership with animal welfare groups and shelters from across the country.

Private groups that one might not expect to be involved after a wildfire also play a key role. The Home Depot Foundation, as an example, was founded in 2011 focusing on causes related to veterans, such as their Veteran Housing Grants Program for the construction or rehabilitation of permanent supportive housing for veterans. In 2025, the foundation committed $3 million to support organizations related to immediate disaster relief and long-term recovery needs in Southern California. The organization partnered with non-profits like Team Rubicon and Convoy of Hope on the immediate post-fire needs, and pledged $1 million to its own Path to Pro program to help fill the skilled labor gap as communities rebuild.

There are also groups, most of them non-profits, that deal with localized wildfire issues. One of the first of these localized groups was created in 1975 to coordinate and improve the wildfire activities of federal agencies in south Florida, where development was rapidly growing in the Miami area adjacent to the Everglades. California has been the leader in the creation of Fire Safe Councils—now more than 300 of them within the state—formed at the grassroots level to make their community less vulnerable to wildfires. The state's efforts are managed by the California Fire Safe Council, established in 1993. In 1989, the North Florida Prescribed Fire Council was formed specifically to focus on prescribed fire. The individual fire councils that grew in other regions and communities became the Coalition of Prescribed Fire Council in 2006.

Other states, especially along the West Coast, are participants in the National Fire Protection Association's Firewise USA. This is a voluntary program that requires participating communities to meet certain criteria involving education and action, a process that often takes six months to a year to implement and involves the entire community and a financial commitment. Fire Safe Councils, in comparison, focus more on what individual homeowners can do to make their residence less vulnerable to wildfires.

Further Reading

Ad Council. 2023. "New Smokey Bear PSAs Encourage Americans to Channel the 'Smokey Within' and Help Prevent Unwanted, Human-Caused Wildfires." September 23. https://www.adcouncil.org/learn-with-us/press-releases/new-smokey-bear-psas-encourage-americans-to-channel-the-smokey-within-prevent-wildfires/.

Brassill, Gillian. 2023. "More Wildfires, More Retardant? Some Question Its Efficacy." *Governing*. April 21. https://www.governing.com/now/more-wildfire-more-retardant-some-question-its-efficacy/.

Congressional Research Service. 2023. "Wildfire Statistics." June 1. https://crsreports.congress.gov/.

Day, Gordon M. 1998. *In Search of New England's Native Past*. Amherst: University of Massachusetts Press.

Egan, Timothy. 2009. *The Big Burn: Teddy Roosevelt and the Fire That Saved America*. Boston, MA: Houghton Mifflin Harcourt.

Hanson, Chad. 2021. *Smokescreen: Debunking Wildfire Myths to Save Our Forests and Our Climate*. Lexington: University Press of Kentucky.

Hawley, Ralph C. 1923. "The Railroad Fire." *Journal of Forestry* 21 (5): 497–501. https://doi.org/10/1093/jof/21.5.497/.

Hayden, Sarah. 2024. "Wildfire Data Is Flawed—Here's How to Fix It." *The Revelator*. October 7. https://therevelator.org/wildfire-data-fix/.

Kaufman, Obi. 2024. *The State of Fire: Why California Burns*. Berkeley, CA: Heydey Publishing.

Kettl, Donald F. 2015. "When It Comes to Wildfires, Collaboration Causes Confusion." *Governing*. September 21. https://www.governing.com/archive/gov-wildfire-paradox.html.

Lada, Brian. 2025. "Wildfire Season in US Could 'Rapidly Escalate' amid Building Heat, Drought." *Accuweather*. April 23. https://www.accuweather.com/en/weather-news/wildfire-season-in-us-could-rapidly-escalate-amid-building-heat-drought/1764913.

Lopez, Steve. 2003. "Using Fires to Blow Political Smoke." *Los Angeles Times*. October 29. https://www.latimes.com/news/local./.

Margolis, Jacob. 2025. "How Much Toxic Heavy Metal Is in That Bright Red Fire Retardant? We Had It Tested to Find Out." *LAist*. April 3. https://laist.com/news/climate-environment/how-much-toxic-heavy-metal-is-in-that-bright-red-fire-retardant-we-had-it-tested-to-find-out/.

Morrison, Ellen Earnhardt. 1976. *Guardian of the Forest: A History of the Smokey Bear Program*. New York: Vantage.

Nowacki, Gregory J., and Marc D. Abrams. 2008. "The Demise of Fire and 'Mesophication' of Forests in the Eastern United States." *BioScience* 58 (2): 123–38. https://doi.org/10.1641/b580207/.

Pyne, Stephen J. 1982. *Fire in America: A Cultural History of Wildland and Rural Fire*. Princeton, NJ: Princeton University Press.

Pyne, Stephen J. 1995. *World Fire: The Culture of Fire on Earth*. New York: Henry Holt.

Pyne, Stephen J. 2023. *The Pyrocene: How We Created an Age of Fire, and What Happens Next*. Berkeley: University of California Press.

Schammel, Marella H., Samantha J. Gold, and Daniel L. McCurry. 2024. "Metals in Wildfire Suppressants." *Environmental Science & Technology*. October 30. https://doi.org/10.1021/acs.estlett.4c00727/.

Stringer, Grant. 2024. "Indictment of US Forest Service 'Burn Boss' in Oregon Could Chill 'Good Fires' across the Country." *Inside Climate News*. February 8. https://insideclimatenews.org/08022024/indictment-of-us-forest-service-burn-boss-in-oregon-could-chill-good-fires-across-country/.

US Forest Service. Department of Agriculture. n.d. "Interagency Wildland Fire Chemicals Policy and Guidance." https://www.fs.usda.gov/managing-land/fire/chemicals/andhttps://www.fs.usda.gov/sites/default/files/2023-Retardant-Use-on-NFS-Lands.pdf/.

US Forest Service. Department of Agriculture. 2023. *National Prescribed Fire Resource Mobilization Strategy*. https://www.fs.usda.gov/sites/default/files/2023-06/Rx-Fire-Strategy.pdf/.

White House. 2025. "Immediate Expansion of American Timber Production." Presidential Actions. March 1. https://www.whitehouse.gov/presidential-actions/2025/03/immediate-expansion-of-american-timber-production/.

Wigglesworth, Alex. 2023. "Lawsuit Jeopardizes Use of Crucial Wildfire Retardant, US Forest Service Claims." *Phys.org*. March 30. https://phys.org/news/2023-03-lawsuit-jeopardizes-crucial-wildfire-retardant.html/.

Wilbur, C. Keith. 1996. *The New England Indians*. Guilford, CT: The Globe Pequot Press.

Williams, Gerald W. 2005. *The USDA Forest Service: The First Century*. Washington, DC: US Department of Agriculture. Forest Service. https://www.fs.usda.gov/sites/default/files/media/2015/06/The_USDA_Forest_Service_TheFirstCentury.pdf.

Wilson, Conrad. 2024. "Criminal Case Dismissed against Federal Burn Manager in Rural Oregon." Oregon Public Broadcasting. June 6. https://www.opb.org/article/2024/06/06/criminal-charge-dismissed-grant-county-oregon-forest-service-prescribed-burn-manager/.

2

Problems, Controversies, and Solutions

Introduction

The terminology used in describing the multiple wildfires that destroyed more than 18,000 structures in communities across the Los Angeles Basin in January 2025 was not hyperbole. Apocalyptic was a common word used to describe the scene; Herculean described the effort that would be needed to restore the neighborhoods and infrastructure in Altadena and Pacific Palisades that cost thirty people their lives, with 200,000 residents forced to evacuate. The estimates of the total economic loss kept climbing, billions at a time. According to a study by the Institute for Applied Economics, total property losses were estimated at $28 billion to $54 billion, with the cost of business disruptions projected to cause $4.6 billion to $8.9 billion in lost economic output in Los Angeles County over the next five years. In addition, employment losses were calculated at between 24,990 and 49,110 job-years with labor income reductions ranging from $1.9 billion to $3.7 billion. Federal, state, and local governments could see tax revenue losses between $0.73 billion and $1.4 billion due to reduced business activity and employment (Institute for Applied Economics, 2025).

Just two months later, out-of-control wildfires burned in central Texas, North and South Carolina, and on New York's Long Island, reminding the public that wildfires were not limited to southern California. Negligence played a role in the fires in North Carolina and New York, including burning debris close to a tree line and a family's attempt to make s'mores on a windy day. Like the wildfires

---

Colorado's 2021 Marshall Fire destroyed 1,100 homes and caused an estimated $2 billion in damage. The state legislature responded by enacting a Fair Access to Insurance Requirements (FAIR) Plan as a safety net for homeowners who were unable to obtain coverage from a traditional carrier and by creating a Wildfire Resilience Code Board charged with developing a statewide wildfire code (Zach Joing/Dreamstime.com).

in Los Angeles, these fires were driven by low humidity and windy conditions, parched vegetation, and in at least one case, downed power lines. Residents were evacuated as multiple fire agencies were deployed to fight back the flames. But the similarities between these examples ended there. None of the March 2025 fires led to dozens of deaths, thousands of burned structures, or catastrophic economic losses. The president did not visit the sites, and media coverage was short-lived and mostly minimal outside the immediate area. Most fires were contained within hours or days. It is against this backdrop that this chapter begins by explaining the myriad impacts of wildfires, from smaller blazes that affect a few acres and are rarely publicized, to the largest blazes that take lives, destroy property, and in many instances, ruin communities.

The Impacts of Wildfires

When wildfire killed 100 people and destroyed much of Lahaina, Hawaii, on August 8, 2023, the damage estimates for property were around $6 billion. More than 2,000 residents were displaced by the fire with limited housing available for those who decided to stay. Many longtime residents could not afford the cost of housing and chose to leave Maui altogether, moves believed to change the island's demographics in the long term. The destruction of the historic downtown devastated the region's economy, leading to a reduction in tax revenue that was estimated to represent 10 to 15 percent of the state's tourism industry.

In southern Oregon, the 2020 Labor Day fires destroyed more than 4,000 homes in communities from Ashland to Medford, becoming the state's most expensive natural disaster. What residents and property owners did not anticipate, however, was that in addition to the hassle of filing claims for their immediate property losses, they would face another hurdle when they decided to rebuild—the cost of reinsurance. Premiums went up an estimated 30 percent, and many homeowners received notices from their insurance companies that their policies had been canceled due to wildfire risk.

California's Dixie Fire in 2021 caused a different result altogether. The fire turned the runoff from Mill Creek black with ash, affecting the spring-run populations of Chinook salmon in the northern Sacramento Valley. In 2023, only thirty-four adults of the species returned to Mill Creek, despite being protected since 1999 by the federal Endangered Species Act. Too much sediment can smother the salmon eggs and baby fish, while large debris can scour the streams and pools where the fish reside. In other areas, the heavy use of fire retardant, toxic to fish, represents a tradeoff between a species on the edge of extirpation and fighting the fire.

These examples of the impact of wildfires represent a type of research that is still in its infancy. Typically, wildfires have been expressed in lives lost, property damaged, or acreage burned. But now different metrics are being used that illustrate that the impact of a fire might be more than just the direct cost that can be estimated, evaluated, and compensated. Now, other impacts are becoming part of the accounting that takes place, sometimes years after the fire has stopped smoldering.

There is no "average" wildfire; there are record-setting wildfires that are differentiated by estimates and numbers. How many people died? How many structures were burned? How much will it cost to replace? How big was it? How many days, weeks, or months did it last? Officials point to the "worst" wildfires within their jurisdiction with all or most of these indices, seldom able to come up with an accurate cost because of the inability to put a price on a life lost, or a viewshed destroyed, or the number of hours a firefighter was away from home. The numbers do tell part of the story, of course. It is possible in some instances to calculate how much a jurisdiction has spent on fighting a single fire in terms of paying for firefighter salaries or support equipment such as air tankers, but harder to measure the loss to families and the larger community, such as when nineteen members of the Granite Mountain Hotshots died fighting the Yarnell Hill Fire near Prescott, Arizona, in 2013. The number of people killed during a wildfire is just one metric for assessing the human loss. In the 1871 Peshtigo Fire in Wisconsin, an estimated 1,200 to 2,400 individuals perished throughout the region. Many of the bodies that had been unrecognizable were buried in an unmarked grave. Two hundred people died together in a tavern; an untold number died jumping off a bridge hoping to escape the flames.

Potentially Compensable Losses: Who Pays?

In April 2022, the largest wildfire in New Mexico's history, the Calf Canyon/Hermits Peak Fire, was spread by two planned burns set by the US Forest Service that destroyed more than 430 homes in the northern part of the state. An estimated 25,000 people were told to evacuate, some more than once until the wildfire was brought under control. Because the fires were started by a federal agency, victims were eligible to file for reimbursement of their losses under the Federal Torts Claims Act. The law limits compensation, however, to tangible items like destroyed property where there is a specific price tag. Even while the flames still burned, two members of the state's Congressional delegation introduced legislation, the Hermits Peak Fire Assistance Act, which was enacted the following September that established a $4 billion compensation fund for the fire's victims. It required FEMA, the Federal Emergency Management Agency, to fully compensate residents and business owners impacted by the blaze. The bill establishing the fund was modeled on the Cerro Grande Fire Assistance Act that was created after a similar fire in New Mexico in 2000. Both measures included language stating that payments would be limited "to actual compensatory damages." However, President Joe Biden promised that the federal government would "fully compensate" the victims for the USFS miscalculation that allowed the fire to get out of control. To many of the victims, "full compensation" meant just that—compensation for any loss attributed to the fire, such as the cost of having to move to a different city or state because a house was destroyed, or the reduction in value of a structure like a vacation home that had a gorgeous view of the forest that is now just burnt stumps. Does the federal law or the New Mexico-specific legislation cover noneconomic losses like displacement when forced to move, or mental health treatment due to the stress of a wildfire?

So, who pays and what is covered? It depends.

When referring to the damages and loss that individuals incur, the answer often begins with home insurance for which the person has paid a premium. People living in areas considered at risk or prone to wildfires are increasingly finding that the cost of paying a home insurance premium has ridden exponentially, assuming they can even find a company to issue a policy. For some, such as renters, or those with what might be called non-traditional housing such as a granny flat or unattached structure, there may be no insurance policy under which a claim could be filed at all.

In other instances, when a person is able to identify the cause of the wildfire or a party is potentially liable for damage, a civil lawsuit may be filed. In these kinds of cases, the damaged party (the plaintiff) might begin by totaling up "actual" damages, which usually refers to the monetary damage incurred in the fire, such as the house that burned, and the contents of that house such as furniture, clothing, and kitchen items. If the individual was employed, they may seek indirect damages such as compensation for lost wages, or noneconomic damages sometimes referred to colloquially as pain and suffering, or emotional distress. In other instances, the party that brings the suit may ask for punitive damages to punish the liable entity.

When a case is heard in court, the suing party can always ask for payment for any one or more of these types of damages, but they may not always succeed. In other situations, the liable party may find that the indirect damages awarded to the plaintiff are higher than the actual damages. *High Country News* found, for example, that a jury in the 2020 Labor Day Fires in Oregon ordered PacifiCorp to pay nine plaintiffs $62 million, although the actual damages amounted to only $6.3 million. The amount awarded may seem unreasonable to some, but others may feel that no dollar figure can adequately pay for someone's losses. In another Labor Day Fires case involving the utility's power lines, a jury found PacifiCorp fully responsible for the Santiam Fire. Altogether, the plaintiffs won more than $73 million, with each one receiving between $3 million and $5.5 million for physical damages and emotional distress (Mohr, 2024).

Not every injured plaintiff can claim success, as seen in what happened to the residents of Grizzly Flats, California, after the August 2021 Caldor Fire burned for sixty-seven days. The fire burned across the Eldorado National Forest in less than fifteen minutes, burning over 200,000 acres and destroying more than 400 of the town's 600 homes. Two men were arrested and charged with reckless arson but pleaded not guilty, and a judge ruled there was insufficient evidence to take the case to trial. In the meantime, local residents complained that the USFS had failed to adequately manage the overgrown lands surrounding the community, resulting in a Congressional investigation and a segment on the television program *60 Minutes*. Some Grizzly Flats residents had not been able to afford the expensive premium for wildfire insurance and could not afford to rebuild after the fire. Several hundred others who lost everything joined together in August 2023 to file claims against the Forest Service for damages under the Federal Tort Claims Act, but their claims were denied or the USFS did not respond to their claim. They hoped a law firm would take on the USFS and would handle their claims; the first seventy-five claims that were being reviewed by one firm sought $839 million in losses.

Although the Federal Torts Claims Act provides the mechanism for those who have incurred damages due to wildfires in some cases, justice is not necessarily swift. Plaintiffs in these cases must be patient and willing to wait years (and years) for the final resolution of their claims. For instance, hundreds of survivors of the Chimney Tops 2 Fire that originated in the Great Smoky Mountains National Park near Gatlinburg, Tennessee, on November 23, 2016, filed suit against the National Park Service (NPS) and the US Department of the Interior. The plaintiffs alleged that wrongful deaths, personal injuries, and property damage occurred because the NPS had failed to control a prescribed fire that breached the park's boundaries, eventually destroying more than 2,500 structures over 17,000 acres and resulting in the deaths of twelve individuals. Working its way through the legal system, the case was still mired in the courts more than eight years later. In northern New Mexico, meanwhile, a San Diego-based law firm represented 1,125 clients from the Hermits Peak Fire seeking noneconomic damage payments from FEMA under New Mexico law. According to an investigation by the non-profit group ProPublica, the law firm anticipated that almost all will qualify for some type of payment.

The question of who pays also depends on what is sought. Just weeks after the January 2025 fires in California, the governor sent a letter to Congress seeking nearly $40 billion to help the Los Angeles area recover, noting "Los Angeles is one of the most economically productive places on the globe, but it can only rebound and flourish with support from the federal government as it recovers from this unprecedented disaster." The governor sought an additional $16.8 billion from FEMA for the rebuilding of property and infrastructure, with $5 billion of the money for debris cleanup. An additional $10 billion was sought for grants to fire victims, homeowners, businesses, and renters from the Department of Housing and Urban Development, and $5.3 billion for homeowner and business loans from the Small Business Administration (Associated Press, 2025).

Such a request is not itself unprecedented. States and communities routinely seek various types of federal assistance after a disaster, and the requests are usually granted. What was different in the California wildfires was the amount of aid sought, and the political environment of the ask. Just before Christmas 2024, Congress approved billions of dollars in relief funds for three previous disasters—wildfires on Maui, Hurricane Helene, and Hurricane Milton. But it took a year and a half after the wildfire in Hawaii, and more than two months after the hurricanes devastated communities in North Carolina and parts of the southeastern United States for Congress to start the process.

The Cost of Fighting a Fire

Suppression costs are those activities aimed at restricting the spread of a wildfire after it has been detected. The cost of halting a wildfire from beginning to end makes up only about 8 percent of the actual costs that can be attributed to the fire itself. It is difficult to put a dollar figure next to the financial costs borne by the agencies that fight fires because there are so many differences in how the costs are calculated. The determinations are calculated and vary considerably from one jurisdiction to another, making comparisons difficult. The University of Idaho's Policy Analysis Group has been successful in deriving some state costs for suppression, but reports indicate many inconsistencies. In the 2022 Western Forestry Leadership Coalition report on wildfire costs, data on suppression costs for nonfederal partners ranges from excellent to nonexistent. As a result, it is difficult to get a complete picture of suppression costs across the country, except for the largest fires. Data for fires within a local responsibility area (LRA) are typically not available since there is no state requirement that these types of data be reported. There is only limited data for incidents that involve both state and federal responsibility areas that occurred prior to 2014.

Statistics related to the suppression costs for federal wildfire efforts are usually considered to be more comprehensive than those of state and local governments, especially the ones from the National Interagency Fire Center (NIFC). The agency compiles the total number of fires with the acreage affected and suppression costs for the USFS and other agencies within the US Department of Agriculture (Bureau of Indian Affairs, Bureau of Land Management, Fish and Wildlife Service, and National Park Service) in a yearly report. In years when the number of acres burned was exceptionally high, as in 2017 and 2015, the costs were correspondingly high (over $1 billion for the USFS). In 2018, when the number of acres burned was much less, the USFS cost was low, too. In 2021, when there were 58,985 fires reported over 7,125,643 acres, the cost totaled over $3.7 billion. Two years later, a similar number of fires (56,580) burned about half as many acres (2,693,910), but the cost was still the second highest ($2.7 billion) of any year during the reporting period 1985–2023. To date, there has been little discussion of why some years have more expensive suppression costs than others, although it is understood that some equipment used in fires, such as air tankers and helicopters, are more expensive line items than others.

Accounting for Impacts

Just two months after the January 2025 wildfires in Los Angeles, researchers at UCLA's Anderson School of Management had already estimated the economic impacts of the series of blazes. While some of the metrics were relatively easy to retrieve, such as the 16,251 homes, commercial buildings, and other structures that were destroyed, or the number of deaths, more difficult to assess were the impacts on other key indicators. For instance, the total direct insured property loss was not immediately available because the California Department of Insurance had not calculated the amount of the insured claims submitted, nor were figures easily estimated for losses that were not covered, such as those properties without mortgages or those underinsured. Often left out of the impacts equation is the fiscal impact on governments, especially the municipalities where wildfires affect sales and property taxes. Data on how wildfires release carbon and other types of pollutants is still somewhat hard to find, although some studies have linked premature deaths to specific economic costs.

The UCLA researchers made preliminary estimates of the percentage of the population in counties exposed to wildfire burned areas, linking that measure to county-level gross domestic product (GDP), then analyzed that number by eleven business sectors. Using that formula, they estimated that the Los Angeles fires will cause about a $4.6 billion loss in the county's GDP in 2025. Other metrics showed that the region's housing market, especially for rental units, will become increasingly unaffordable (Li and Yu, 2025). These complex calculations may not be appropriate for wildfires on a smaller scale, however, or for affected communities with much smaller populations or a less diverse economic base. But this type of post-fire accounting is important for policymakers attempting to determine future spending priorities and levels, for planning purposes, and for determining whether mitigation is necessary. For the gateway communities surrounding Yellowstone and Yosemite National Parks, for instance, the impact of wildfires on the region's tourism may be of key importance, while in more urbanized areas like New Jersey, analysts will look more closely at the impact of a wildfire on human health.

Natural Resources

Unlike landscapes such as grasslands or rangelands that regenerate and grow back relatively quickly, stands of timber can take generations to return to their original economic value. For this reason, policy emphasis (as well as allocation of human resources) was traditionally on forest fires. Landscapes in states like California are not recovering from wildfires the way they used to or the way researchers might expect. One reason may be the severe drought conditions that have plagued western states for the past two decades. As climate change continues, it affects vegetation by reducing plant production. The regeneration curve, as it is known, lags behind as the ecosystem attempts to recover. Seedlings have a more difficult time trying to survive during a massive drought. Trees dry out and end up dying.

Wildlife and their habitats can be affected in both positive and negative ways by wildfire. On the plus side, fires can create changes that benefit birds and mammals such as downed trees or new meadows. Researchers have noted that the increasing scale and severity of fires make them harder on some species than others. The megafires that scorched California's Western Cordillera ecoregion in 2020 and 2021 burned more than 4.6 million acres and habitat for more than 500 vertebrate species. More than 100 species experienced high-severity fire (defined as one that leaves behind at least 70 to 80 percent mortality in overstory trees) that burned between 5 and 14 percent of their range in one study. Among those species is the long-toed salamander, estimated to have had 14 percent of its range burned, the most of any species (Pope, 2024).

Researchers are now identifying a variety of wildfire impacts that affect water quality, since nearly two-thirds of the United States' drinking water comes from forested areas. In the case of the 2025 fires in Los Angeles, the State Water Board issued a formal order requiring screening for benzene contamination in the areas impacted by the fires, followed by the issuance of a public notification of a drinking water advisory that affected nine public drinking water systems. The do-not-drink/do-not-boil advisory was not lifted until four months after the fires, and efforts at restoring water infrastructure and upgrading system resiliency continued for months after that. In another example, after the residents of Las Vegas, New Mexico, survived the 2022 Calf Canyon/Hermits Peak Fire, they were warned by the city's mayor that there was another challenge looming: the city of 13,000 people had only thirty days of clean water left. The 2020 monsoon rains had covered the region's watershed where the high-severity fires had burned through over 340,000 acres of trees and vegetation in and near the Santa Fe National Forest. The fires introduced massive amounts of carbon and other contaminants into local streams and reservoirs, and water treatment plants could not keep up, making their stores of water undrinkable.

Another growing area of research relates to the impact of wildfire smoke on natural resources. Ecologists from the University of California, Davis Tahoe Environmental Research Center found that smoke can alter lake health by affecting components such as light, temperature, and oxygen levels in the water. Although the exact impact varies, the dense smoke and ash from wildfires are known to disrupt the amount of sunlight that reaches a lake's plant and aquatic life. Burned areas fundamentally alter a watershed's hydrology. As wildfires burn hotter, consuming more trees and structures, water quality will continue to worsen. In another study conducted by scientists at the University of California, Santa Cruz, the researchers found that wildfires can change the chemistry of nearby streams that are the sources for both wildlife and humans. While more information is needed before they can determine what impact these changes in drinking water have on human health, there is agreement that routine water-chemistry monitoring is needed.

Coupled with the damage to soil are the dangers related to erosion. Heavy rainfall in Montecito, California, after the 2018 Thomas Fire led to an avalanche of mud, trees, and other debris that destroyed 130 homes and killed twenty-three people. Similar incidents involving debris flows after wildfires have led officials to think of fires not just as flames but also as mudslides that can affect infrastructure years after the wildfire is over. After the 2025 Los Angeles fires, local officials issued evacuation warnings for residents in the areas that had burned along with nearby communities as storms hit the area. Fire officials warned that debris flows could fill canyons and low-lying areas with fast-moving water, potentially making it difficult for emergency personnel to access affected zones, trapping individuals who failed to leave in time. Current efforts to control debris flows, such as mountain catchments, debris basins, and flood control channels may be insufficient in protecting public safety and causing collateral damage even in areas that wildfires do not burn.

Human Health

Research into the health effects of wildfires is growing, but the results show that there is much more to be done to determine the nature of the risk of exposure. Some components, like those in smoke, are considered of particular concern. Termed an overlooked hazard by some researchers, airborne hexavalent chromium, also known as chromium 6, can be transformed by wildfires into ultra-fine dust that can be inhaled, ingested in drinking water, or carried downstream. Wildfire-modified chromium 6 increases cancer risk in ways that are different from the benign form of the metal commonly found in soils and plants. The primary risk is for people working to rebuild previously burned areas, or recreationists wanting to check out a burn scar while hiking.

Past research has clearly identified the role of ambient air pollutants in causing health problems such as asthma and cardiovascular disease, but there are new studies that mirror those findings when the pollutant is wildfire smoke. A decade-long study funded by the US National Institute on Aging found that exposure to wildfire smoke increases the risk of being diagnosed with dementia more than other forms of air pollution. The researchers studied 1.2 million southern Californians over age sixty over a ten-year period to determine the effect of exposure to fine particulate matter based on where they lived. The pollutant released from wildfires was found to have a higher concentration of toxic chemicals and smaller than the fine particulate matter from other sources. The findings were more pronounced, the study found, in individuals from racial and ethnic minoritized groups and in high poverty areas (Scripps News Staff, 2024).

Researchers have found that there is evidence that particulate matter from wildfire smoke is more harmful to human health than the particulate matter that comes from other pollution sources, a finding that has special importance in a region like southern California, which has both ongoing wildfires and widespread poor air quality. One UCLA study published in 2024 found that thousands more people die due to the health impacts of wildfire smoke in California than previously counted. By inhaling the fine particulate matter in smoke from wildland fires, the study found that an estimated 52,500 to 55,700 deaths had occurred, with an associated economic impact of $432 billion to $456 billion (University of California, 2024). Since that time, there have been several record-setting fires, including the ones from January 2025, which will lead to significant health burdens. Those impacts occur even in areas not prone to wildland fires, such as those in 2024 in New Jersey and New York, that produced massive amounts of smoke. While there is a health risk from air pollutants in the immediate fire zone, the risk extends to those even as far as 150 miles away, as was the case in the 2018 Camp Fire in Butte County, California.

Existing measures of wildland smoke are not found in the Air Quality Index (AQI), which excludes large-particle pollution like ash and gaseous volatile organic compounds resulting from a fire. The smoke from the multiple blazes that burned in the Los Angeles Basin in 2025 is different from what is found in smoke from a forest fire. In Los Angeles, the clouds of smoke included everything that burned, from vehicles and electronics to all types of metals and plastics. The urban component of the fires leads to smoke that includes lead, cadmium, and aluminum, creating what leaders called an unprecedented public health crisis. Exposure to urban wildfire pollution can result in both acute and chronic health impacts, especially for those with cardiovascular issues or compromised respiratory systems, with more research needed on the potential long-term effects. As a result of these warnings, the LA Department of Public Health prohibited the removal of fire debris from affected areas until agencies performed hazardous materials inspections. Even in areas that did not burn, the health risks from smoke are significant, causing the adjacent city of Pasadena to declare a local health emergency. It is unknown how long these effects will linger, but the publication *Inside Climate News* cited a 2024 preliminary case study that suggested wildfire air pollution can land and stay trapped in urban basins far away from the fires (Gray and Robinson, 2025). Shortly after the fires started, a team of scientists from the California Institute of Technology found massive spikes in lead and other toxic chemicals at a monitoring site in Pico Rivera, more than 10 miles from the Eaton Fire. Pollutants can be stirred up and remixed by the wind or by resuspended by the cleanup and debris removal process and can settle into soil and dust.

These impacts are being addressed in a ten-year project launched by a team of university researchers called the Los Angeles Fire Human Exposure and Long-Term Health Study, or LA Fire HEALTH Study. The longitudinal research began testing the air, soil, water, and dust in and around homes at fifteen monitoring locations throughout Los Angeles County after the January 2025 Palisades and Eaton Fires. The team, made up of researchers from Harvard, UCLA, the University of California, Davis, and the University of Texas, will also be deploying specially-equipped vans through the affected neighborhoods to measure the chemical composition of particulate matter and gases on each block in real time. Unlike the AQI, the data will be much more detailed in identifying which pollutants are found in particulate matter, and in what amounts. The researchers will also use advanced modeling techniques and equipment to determine and predict how the smoke from the two wildfires may have penetrated the 14 million homes in the LA region, which is shaped like a bowl surrounded by mountains. Some pollutants may have been trapped by buildings, infiltrating structures not immediately adjacent to the fire areas. Included in that part of the study will be research into how the smoke and other hazardous compounds might continually be released from porous materials such as carpeting and furniture. The long-term implications of this study, along with others in the planning and experimental stage, will potentially aid in determining the health impacts of wildfires on a scale that has never been done before.

Ecosystems and Biodiversity

Wildfires can have a tremendous effect on ecosystems, biodiversity, and aquatic health. In recent years, wildfire smoke covered as much as 70 percent of California's land area, including freshwater ecosystems, according to a study by Dr. Steven Sadro of the University of California, Davis. "We're looking at a scenario where for the next 100 years or longer, smoke will be a feature on the landscape" (Udasin, 2024). But it is difficult to collect post-wildfire water quality data because the effects vary over time. The most severe impacts often do not appear until there are intense rainstorms, phenomena that may not happen immediately after the wildfire, and may not happen until months or years have transpired. By then, data collection opportunities (and funding) may have vanished, and it becomes impossible to determine how the loss of canopy vegetation, soil changes, and erosion after the fire affect the water and species in an ecosystem.

The wildfires that burned across the Texas Panhandle in 2024 were unprecedented in many ways, including the impact on wildlife habitat and resiliency. Unlike forest fires more typical in the West, much of the 900,000 acres burned in the Smokehouse Creek Fire was composed of rangeland and mixed cottonwood and tallgrass bottomland ecosystems. Despite the number of acres that burned, wildlife specialists were optimistic about the ability of both the land and its wildlife to rebound. Noting the relationship between wildlife and vegetation, Dr. Jacob Dykes of Texas A&M University said that the two have evolved alongside for millennia. Mammals such as pronghorn can see and smell an approaching fire with time to run and escape, he said, and birds can fly to an area that is safe, while burrowing animals can retreat to their dens underground. The wildlife at risk were more likely to be smaller ones that are older or less mobile (Fuller, 2024).

The timing of a wildfire also affects the ability of wildlife and vegetation to rebound. In Texas, the blazes occurred prior to the nesting season of ground nesting birds, and the fawning season for deer had not yet started. In one Wildlife Management Area along the Canadian River studied by researchers, there were pockets of unburned rangeland that provided a safe refuge for species like turkey and quail. Fires can actually be seen as a fertilization event for vegetation, Dykes said, breaking down nutrients and delivering them to the soil. As the grasses and flowering plants return, the new vegetation will increase wildlife cover and contribute to species diversity. Studies have shown that following a fire, the crude protein content in tallgrass prairie forage is up to four times greater than forage in areas that have not burned, benefitting both wildlife and livestock.

Recreation and Tourism

An app called TheDyrt.com is a source for many outdoor enthusiasts looking for information about camping availability, reviews, photographs, and tips as they make travel plans. In 2024, the site noted that 18 percent of campers reported that wildfires or natural disasters disrupted their camping plans in 2023, three times more than the 6 percent who said their plans were changed in 2019. In 2020, the number was 11 percent, and then 18 percent in 2022. Three states were found to be the most affected—California, Oregon, and Washington. The Dyrt said that because it can be difficult to get a reservation at popular campsites, making a last-minute change of plans due to a wildfire is exceedingly difficult. Even after a wildfire is over, a campground can be closed indefinitely as trucks and equipment clear away debris like burnt trees and structures, officials make plans on how to rebuild and restore a site, and infrastructure repairs and improvements are made. In times of financial constraints, public sites may be closed indefinitely, funneling campers to privately owned recreation areas or cancellations of tours and trips altogether.

When wildfire spreads to recreation areas, the damage may take decades, rather than months or years, to repair. In other locations, buildings can be replaced, downed trees can be removed, and trails can be rebuilt. For the three California state parks that burned in 2020 in the CZU Lightning Complex Fire, fully reopening to visitors has required years of navigating through the state's complex system of permits, studies, and regulations. The area is home to old-growth redwood trees and endangered marbled murrelets that limit construction. Over 24,000 acres burned, destroying the visitor center, museum, and store in Big Basin, California's oldest state park. Flames reached over 300 feet high, altering the forest and leading to a significant loss of tree canopy. Utility and water treatment systems were damaged, closing campgrounds and restrooms, and changing, at least to some extent, the visitor experience.

But officials note that rebuilding can also mean recovery and resilience. The Save the Redwoods League and California State Parks partnered to protect the three parks with an October 2024 Forest Management Strategy that called for prescribed burns to mimic natural fire patterns and thinning dense forests to promote the growth of larger trees. Clearing overly dense vegetation and thinning unhealthy trees will allow the redwoods to withstand wildfires in the future. Some trails needed to be rebuilt, but with an eye toward restoring the natural hydrology that provides more water for the thirsty redwoods. A Facilities Management Plan has allowed park planners to revise building footprints and parking areas to improve the overall operation and resilience of the parks.

Sometimes even the risk of a fire can affect a tourism attraction or place of interest, and there may be little notice to let visitors know. Hawai'i Volcanoes National Park is one such example. In July 2024, park officials closed sections of roads to motorized vehicles even though no wildfire was present, prohibiting open fires at a campground, picnic area, and at the Kilauea Military Camp. They cited low rainfall, reduced humidity, and gusty winds that they said had created dry conditions through the national park, warning visitors not to park on dry grasses. In a notice reminiscent of Smokey Bear, the park stated, "Every visitor has a role to help prevent wildfires in national parks."

Los Angeles faced different impacts from the January 2025 fires, some monumental, others perhaps somewhat trivial. The suite of blazes led to an entertainment production shutdown across the basin, which had an unknown economic impact as everyone from actors to craft services personnel was placed on indefinite leave. Whether out of necessity (a lack of personnel) or scruples, the Critics Choice Awards, scheduled to be held in nearby Santa Monica, were postponed for nearly a month. Film and TV industry screenings, luncheons, and gala events were cancelled, filming stopped due to smoke and high winds, and a professional football game was moved from LA to Glendale, Arizona. Fires that broke out in the Hollywood Hills led to evacuations and stopped the kitschy tours of celebrity homes and the Dolby Theater where the Oscar awards are held, closing the Hollywood Walk of Fame to tourists. Even a high-ranking political figure found that his ride, Air Force One, was grounded due to high winds fanned by the flames.

The publication *Tourism Review* provided its readers with a rundown on popular attractions, noting that places like Disneyland and Universal Studios Hollywood were still open but noted that smoke and poor air quality would impact the visitor experience. Other popular tourist sites like the Getty Villa,

Los Angeles Zoo, and Griffith Observatory were temporarily closed. Even routes that draw drivers to test their skills to see the local scenery were closed, such as portions of Pacific Coast Highway and Mulholland Drive. British tourists, on the other hand, were urged to visit Los Angeles as a way of supporting the region's wildfire recovery.

Economic and Business Impacts

The economic impacts of wildfires are somewhat easier to discern, but cover more segments of the economy than one might imagine, both immediately and long past the date a wildfire is contained. Businesses may need to shut down for a short period or they might be damaged so much that they can never reopen. Workers may lose their jobs because of what happened to the business where they worked. Sometimes the economic impacts are not immediately apparent, such as the products affected by the fire and its smoke. Farmers in the Central Valley of California, which produces a majority of the world's almond and walnuts and nearly all of the pistachios sold in the United States, have dealt with the effects of wildfire on the state's important nut crop. A University of California, Davis study after the 2020 fire season in the state found that nut trees produced a significantly smaller crop—as much as 15 percent fewer nuts—the year following the fires. In some nut orchards, the yield was cut in half. While the scattered light from wildfire smoke helps trees photosynthesize and increase their energy storage, the smoke from the 2020 fires was so thick that it decreased the overall light to the trees, leading to a smaller crop. Longer, more intense fires, and their impacts on the nut crop, are "collateral damage" as one researcher called it (Tidmarsh, 2024).

The impact of wildfires on the economy is often calculated in property loss—buildings and inventory, for instance—but there are other types of costs that affect different business segments. The February 2024 fires in the Texas Panhandle, the largest in the state's history, led to $123 million in agricultural losses according to Texas A&M researchers. The fires that burned across the state's rangelands killed an estimated 12,000 cattle at a cost of $27 million, with other associated costs, such as $1 million to dispose of deceased animals. Infrastructure costs of $68.7 million ranged from the repair and replacement of fencing along with stocks of hay or feed, with another $26 million in losses due to damaged pastures and range, some of which required short-term emergency feeding (Fannin, 2024). For ranchers in Texas, grassland fires became exceptionally important to their bottom line.

The wildfire headlines in January 2025 initially focused on the wealthy and celebrities who lost their homes, with much less reported about the thousands of low-income and immigrant workers who also lost their jobs. Many of those whose homes were destroyed worked in the service industry, and many were people of color. According to the UCLA Latino Policy and Politics Institute, Latinos made up just 23 percent of the residents in Pacific Palisades, Altadena, and Pasadena, but they made up 36 percent of all workers there. They had jobs in landscaping, housecleaning, in restaurants, and as domestic workers, and as had been the case after the 2018 Woolsey Fire in Malibu, California, many were likely to lose those jobs permanently. Those who were undocumented do not qualify for federal employment or disaster aid that US citizens can apply for and faced deportation under the Trump administration (Reyes-Velarde and Kuang, 2025).

Other economic disruptions from wildfires may be of both short and longer duration. Initially, for instance, medical services may be stretched thin, hospitals close to a fire area may be turned into command centers, and road closures may hamper the ability of medical workers to get to their jobs, affecting staffing levels and hindering patients needing transport. Power outages are likely to occur at some health facilities, forcing them to rely on generators. Prices may go up for those looking for rental housing after a fire, and availability is likely to be less. Shortages of some supplies may occur for a short time, while other items become hard to find altogether.

Arizona State University professor Hitendra Chaturvedi explained what the Los Angeles area was likely to experience in terms of rebuilding with a major impact on the supply chain like a line of dominoes on the construction and home furnishings industries. The process is likely to take more time than people anticipate, if experience is any guide. The imposition of tariffs on products like Canadian lumber, for example, was expected to hit hard on southern California building even before the wildfires, as were those on many other imported products (Terrill, 2025). The tariffs added a layer of uncertainty to homeowners hoping to rebuild as the Trump administration's policies kept changing from one week to another.

The first rebuilding permit in the Los Angeles fires was issued fifty-seven days after the fires began, faster than permitting in previous incidents. Under the mayor's executive order, the review process for permits was taking about six days. Still, nearly four months after the fires, only twenty-three permits had been issued to rebuild homes damaged or destroyed in Pacific Palisades, despite the fact that over 5,600 structures in the area had been significantly damaged. The uncertainty factor, more than procedures, appears to have hit homeowners, construction firms, architects, and other especially hard. The permitting process, added to soaring costs for labor and materials such as lumber, copper, and roofing, caused several previous homeowners to change their minds, one contractor in Altadena said. Another business owner said that before deals for window and door products could be finalized, vendors walked back their commitments due to uncertainty over tariffs. Most companies cannot afford to stockpile materials due to storage costs (Dwyer, 2025).

The Los Angeles area fires were different from a typical wildland fire because what may have started in heavy vegetation quickly moved to structures, and the damage done was totally different. Consumer products like televisions and computers burned along with vehicles, garages full of paint and solvents, and roofs with solar panels. The result was an urban toxic soup that produced fumes, fire retardant, PHS (particularly hazardous substances), heavy metals, and other contaminants that eventually get washed into the sediment and waterways.

The 2025 California fires dramatically revealed that the impact on the economy and business is different when an urban area is destroyed. Wildfires can destroy a city's infrastructure, including power, water, and telephone lines, pipelines, sewage facilities, roads, and bridges. Those segments of the economy are frequently overlooked, making it more difficult to calculate the real significance of hazard risk, vulnerability, and resilience. The historical data on damage done to infrastructure is almost impossible to capture since there is rarely any documentation available prior to 2000. This is especially true in the wildland-urban interface where the built environment abuts the natural environment. When a vehicle is abandoned on the roadway, as was seen in Lahaina and Pacific Palisades, the burning car or truck damages the concrete and creates spalling that can remove chunks of the road. There are other hazards to consider, such as trees that teeter or fall onto the roadway, structural damage to signs, guardrails, and other elements that continue to exist even after the wildfire has been contained.

Homeowners Insurance

Insurance serves as a defense against the losses people face when a disaster destroys a structure and other personal property. The coverage is in exchange for regularly scheduled premiums that are paid to an insurance company based on the value of what is insured to protect against future loss. The insurance industry assumes the risk of the loss happening and shares the risk by redistributing it among others it covers as its customers. Banks also require insurance in order for someone to obtain a mortgage loan on a home. Insurers sell $15 trillion worth of homeowners insurance to an estimated 85 percent of all US homeowners, who collectively pay $120 billion in premiums. Insurance is regulated at the state level where regulatory agencies review a company's proposals for rate increases based on the loss model they submit, among other factors. Not all states regulate equally, however, and not all policies provide the same kind of insurance.

That was the lesson learned in Lahaina, Hawaii, after the 2023 fires on the island of Maui. Even though some residents carried homeowners insurance, which in some cases covered a provision for loss of use (LOU) if their home was determined to be uninhabitable and additional living expense (ALE) benefits for twelve months, they found that once that period was over, their insurance companies did not extend their benefits, as some had done in other states. In order to apply for rental assistance from FEMA, they had to exhaust their LOU and ALE benefits. For some families, there was a lengthy wait between the application and approval, which meant they had to pay for the mortgage on their destroyed house as well as the cost of rent. The longer they had to pay rent while they struggled with the red tape and time needed to get building permits, architectural plans, contractors, and materials, the less money they had to rebuild. In the first year after the fires, about half moved out of West Maui, and nearly a tenth of that group left for the mainland, according to a 2024 University of Hawaii Economic Research Organization survey (Machado, 2024).

One of the most frequently asked questions posed by homeowners is why their premiums are increasing even though they live in areas where wildfires are less likely to occur and the loss rate is low. The simple answer is that even if a homeowner does not live in a high-risk area, regardless of the type of disaster, that owner is likely paying a higher premium to cover people in the areas where the risk is high. How much rates have increased is debatable. According to the industry's non-profit Insurance Information Association, there was a 10 to 12 percent increase in homeowner insurance costs in 2023, although the National Bureau of Economic Research says the average property insurance premium has increased by 30 percent since 2020 (Baylis, 2021). The American Association of Retired Persons reported a 15 percent increase nationwide from 2021 to 2023, citing several reasons why home insurance rates are rising:

- Home repair costs for supplies such as lumber and roofing materials increased nationwide
- Labor costs, such as hourly earnings, adjusted for inflation, rose 4.6 percent in 2022
- Housing prices have risen 8.9 percent a year for the past five years, meaning the cost of replacing them has also increased
- Short-term interest rates have increased to combat inflation, meaning that insurance companies need to borrow more money to do business
- Reinsurance costs (insurance for insurance companies) premiums have increased, which means homeowners premiums are likely to rise as well (Waggoner, 2023).

Some observers point to climate change as an additional reason for rising home insurance rates, noting that extreme weather events are occurring more frequently, and recent statistics from the National Centers for Environmental Information (NCEI—under the NOAA umbrella) bear this out. In 2024, there were twenty-seven confirmed weather/climate disaster events with losses exceeding $1 billion each in the United States, including one drought event, one flooding event, seventeen severe storm events, five tropical cyclone events, one wildfire event, and two winter storm events. From 1980 to 2024, the annual average is 9.0 events (CPI-adjusted); the annual average for the most recent five years (2020–4) is 23.0 events (National Centers for Environmental Information, 2025).

A survey of more than 100 insurance companies compiled by Nerd Wallet helps explain the differences in the cost of home insurance in a different way by using as a sample the policy for a forty-year-old homeowner with good credit with $300,000 worth of dwelling coverage, $300,000 worth of liability coverage, and a $1,000 deductible. Their analysis of home insurance costs for 2025 found the average cost nationwide to be about $1,915 a year, with State Farm as the cheapest widely available home insurance company, at an average annual rate of $1,935. The figures vary however, by the age and size of the home, the coverage of the policy, by the policyowner's claims history, by their credit rating, and by their location. Some states, for instance, prohibit insurers from using credit history in setting the cost of a policy. Not all insurance companies can offer home insurance in all fifty states. Generally, the cost of insurance for newer homes is less than that for older homes (Schlichter, 2025).

There is a tremendous disparity in home insurance costs when the states are ranked in order for the average rate in 2025. Even though rates jumped about 10 percent between 2024 and 2025, the least expensive state, for instance, is Hawaii, with the cost per year at $610, followed by Vermont ($950), Delaware ($1,025), Alaska ($1,035), and Maine ($1,180) in the top five. Compare that with the five most expensive states and the difference is remarkable. Oklahoma's average cost for homeowners insurance in 2025 was $6,210—ten times that of Hawaii, followed by Texas ($4,585), Nebraska ($4,505), Colorado ($4,1075), and Kansas ($3,737) rounding out the top five (Schlichter, 2025). Interestingly, when the five states with the most expensive home insurance premiums in 2025 are matched against the NCEI map of the twenty-seven disaster events of 2024 with losses exceeding $1 billion, none of them involved wildfires. When looking at a longer period from 1980 to 2024, only 23 of the 403 disaster events that cost $1 billion or more involved wildfires, ranking sixth behind severe storms, tropical cyclones, flooding, drought, and winter storms. In terms of total costs, wildfires for that period ranked fifth, with tropical cyclones more than half the total costs of disasters. Louisiana's residents were among those hit hard by natural disasters, according to a 2023 study by Louisiana State University that found 17 percent of the state's homeowners reported their insurance company had canceled their policy, and 63 percent said the cost of their homeowners insurance had increased from the prior year. Similarly in Florida, Farmers Insurance told policyholders it was too risky for the company to continue insuring their homes, pulling out of the market there entirely. In both states, it was clearly not because of wildfires.

Does this mean that wildfires are not responsible for the high (and increasing) costs of home insurance premiums? Not necessarily. A December 2022 working paper prepared under the auspices of Harvard Business School found that insurance companies that were more regulated by states adjusted their rates less frequently and by a lower magnitude after experiencing losses. To compensate, they adjusted their rates in states where they were less regulated, basically cross-subsidizing across states (Oh et al., 2024). The NCEI maps and data may corroborate that finding on a simplistic level. The distribution of damage from US billion-dollar disaster events from 1980 to 2024 is dominated by tropical cyclone losses. Tropical cyclones have caused the most damage

($1,543.2 billion, CPI-adjusted) and also have the highest average event cost ($23.0 billion per event, CPI-adjusted). A study published as a National Bureau of Economic Research working paper found what the authors called "information asymmetries" in how California's insurers were pricing risk, with some doing so at a very coarse level, such as properties in the same ZIP code, while others priced risk at a granular level, even down to property-level data.

Insurers are responding in a number of ways to the increasingly expensive losses they have faced. After the 2018 Woolsey Fire burned nearly 100,000 acres in Ventura and Los Angeles counties, two companies—Chubb and AIG—stopped insuring homes in fire-prone areas. In March 2024, State Farm, which insured more homes than any other company in California, told 72,000 of its customers, both homeowners and commercial apartment owners, that their policies would not be renewed, representing just over 2 percent of the policies it issued in the state, citing inflation, catastrophe exposure, and reinsurance costs. About 1,600 of the policies that were not renewed were in Pacific Palisades, where one of the 2025 fires started. Two months later, the company said it was pausing issuing any policies in California due to wildfire risks. Other insurance companies decided to leave the state entirely. Part of their rationale is that the cost of rebuilding and replacing a home has skyrocketed. California's median home price rose 5 percent in 2024 to $861,000, more than twice the national median. Los Angeles County's median home price exceeded $900,000, with an average mortgage payment of $5,500 per month including taxes and insurance, which meant that only 11 percent of households could afford to purchase a home (Mendoza, 2025). But as many as 6 million homeowners, or 7.4 percent of all homeowners in the United States, have chosen not to get homeowners insurance—colloquially referred to as "going naked"—which the Consumers Federation of America estimates at about $1.6 trillion of unprotected home value (Meyersohn and Bahney, 2024).

Worse yet for many homeowners, typical policies do not cover the losses from disasters like wildfires unless the person gets a separate policy, called a rider, at an additional expense. Oregon law requires insurers to include wildfire coverage as part of its homeowner policies, but the statutes do not preclude insurance companies from deciding not to renew or insure a policy because of wildfire risk.

Several states have attempted to assist homeowners who are unable to find an insurance company willing to take the risk of providing them with insurance in the traditional marketplace. In California, the nation's fourth largest insurance market, the Fair Access to Insurance Requirements Plan (CalFAIR Plan) is the safety net for homeowners who are unable to get insurance from any of the companies licensed to insure property in the state. The plan is not a state or public entity and is considered a temporary solution until homeowners can obtain coverage from a traditional carrier. Under the FAIR Plan, consumers get wildfire coverage—less coverage—at higher premiums, which is why many refer to it as the insurance company of last resort. The plan has a cap of $3 million for rebuilding, lost possessions, and temporary housing. This means that those with more expensive homes need to cobble together multiple policies to cover potential losses. In 2018 the program had 126,709 residential policies in force (PIF), soaring to 573,730 in March 2025, a 23 percent increase since September 2024 and a 139 percent increase since September 2021. The plan's exposure in March 2025 was $599 billion. This represents a 31 percent increase over the prior fiscal year, September 2024, and a 259 percent increase since the end of the fiscal year in September 2021 (California FAIR Plan, 2025).

The FAIR Plan's New Business (NB) has also increased in the first half of fiscal year 2025 (October 2024 to March 2025). An additional 138,868 dwelling and commercial NB was gained during that period, which represents a 30 percent increase in monthly average NB over FY 2024 and 324 percent increase over the end of FY 2021 (California FAIR Plan, 2025). Residents of Pacific Palisades, where thousands of expensive structures burned, held 85 percent more FAIR Plan policies in September 2024 than they had a year prior. According to the real estate service Redfin, the median price of a Pacific Palisades home was $4 million just before the fires, up 23.4 percent from 2024, and $4.2 million in April 2025 despite the fires (Redfin, 2025).

Some homeowners in Pacific Palisades were able to obtain building permits on their preexisting parcels as early as April 2025 because they hired contractors to clear away the fire debris and decided to rebuild on the same footprint. Many former residents, however, decided to leave because they did not want to live in an area that was likely to be under construction for years. Real estate developers Gatsby Development noted that there were other considerations that could draw out the rebuilding process. In both Altadena and Pacific Palisades, there was a need to restore the community's infrastructure, from roads and sidewalks to utility systems to local amenities like schools, libraries, and hospitals. This massive undertaking, the developer said, requires both skilled workers and ample materials, which might be made more difficult under the Trump administration's policies. With over 40 percent of the construction workforce in Los Angeles being foreign-born, changing edicts on birthright citizenship, residency, and mass deportations could reduce the labor force. Tariff turnarounds and subsequent trade wars began to make materials more expensive and less available, causing construction delays (Gatsby Investment, 2025).

Florida's Citizens Property Insurance Corporation, similar to the FAIR Plan but funded by a surcharge on policyholders, had more than 1.3 million policies in 2024, an increase of about 50 percent over 2023, having a market share of 16 percent, more than any national insurer in the entire state. The number of claims filed raises concerns about solvency as the plans grow in size and cover policyholders in high-risk areas. The US Senate Budget Committee even launched an investigation in 2024 to determine whether Florida's plan had enough funds to withstand future disasters. The California Department of Insurance issued a new regulation in December 2024 that allowed insurance companies to pass the cost of reinsurance on to consumers, with the caveat that the amount cannot exceed an industry standard. Up until that point, California had been the only state that did not allow reinsurance costs to be passed on. New regulations also allowed companies to use catastrophe models that reflect the likelihood of wildfires to raise their homeowner policy rates. The models must also take into account the measures that a property owner has taken, such as home hardening to reduce wildfire damage, a metric that was not in place prior to the recent changes. In exchange, insurance firms must meet a quota of policies in areas deemed disaster-prone.

California's catastrophic fires led to an unprecedented bailout approved by the state's insurance department when the insured losses estimate reached $75 billion. The new state law, passed just a few weeks earlier, allowed the FAIR Plan to draw half of its immediate shortfall from primary insurers to recoup its costs from policyholders to remain solvent. This meant home and fire insurance customers would see the temporary fee added to their premiums, an action that consumer advocacy group Consumer Watchdog vowed to fight in court. The average home insurance cost in the state was $65 per month, with the assessment expected to add at least $60 per household to the 8 million Californians with fire coverage, according to the newsletter, *Mortgage Professional* (Mendoza, 2025). The last time that the state had approved additional funds for the FAIR Plan was in 1993 after fires in Altadena, Malibu, and Topanga—areas that burned again in 2025.

The impact of the assessment was not felt evenly. All private insurers in the state must contribute toward the FAIR Plan's financial stability, which means that homeowners in areas where wildfires are likely to occur would see their premiums rise to the point that they may be forced to participate in the FAIR

Plan due to reduced coverage availability or affordability. With a 2.5-million-unit housing shortage before the fires, rising insurance costs made homeownership even more challenging. Colorado's FAIR Plan, enacted after the 2021 Marshall Fire destroyed 1,100 homes, was a legislative response to an estimated $2 billion in damage. Louisiana and Florida have similar state-affiliated plans in place as a way of filling in the gap due to disasters, although in those states, the cause was not due to wildfires. Insurers also want homeowners to take responsibility for preventable peril.

Debris Removal

The issue of removing the debris after a disaster became abundantly clear in 2005 when Hurricane Katrina destroyed more than 23,000 homes in the New Orleans area alone. The year-long, $2 billion effort involved numerous agencies and organizations across several states, becoming the largest waste disposal project in US history and dwarfing the debris from the collapse of the Twin Trade Towers in 2001. Because landfill space was limited, waste was diverted through a variety of methods, from recycling to burning to grinding. Homes that had to be torn down were demolished in four-square-block sections, with monitoring for potential hazards.

Hurricane debris is different from wildfire debris, though. Wildfire debris usually consists of timber (some of which can be salvageable), with less of a time element involved and fewer concerns about hazards, especially toxic waste. Communities have varied in their approaches when homes and other types of buildings are involved. In Maui's 2023 wildfire, all of the ash and debris from the 1,399 burned residential and 159 commercial lots were cleared by the Army Corps of Engineers (ACE) by February 14, 2025. A temporary site for debris storage was set up in West Maui, with FEMA tasked with the removal of waste to the Central Maui Landfill, the site selected for the permanent disposal site. In California's Woolsey Fire, which burned 97,000 acres and 2,000 structures in 2018, and the Camp Fire, which burned 14,000 structures, debris cleanup was started and completed relatively quickly. Officials said the debris removed from the Woolsey Fire was about 268,000 tons over a period of about a year, with 3.66 million tons removed from the Camp Fire in nine months. In the two major 2025 California fires, where 9,400 structures were destroyed in the Eaton Fire in Altadena and another 6,800 in Pacific Palisades, the situation was different. The sheer volume of debris cleared—about 4.5

million tons—was more than ten times that of the Maui cleanup, and about one ton more than from the Camp Fire.

The process in Los Angeles was complicated by the sheer tonnage of debris and where to put it all. California's governor vowed to complete the debris removal project at "unprecedented, record-breaking speed." The two-phase process began with contractors hired by the EPA sorting through the rubble to locate and dispose of potentially hazardous waste that could not be sent to regular landfills, such as oil, paint, and propane tanks, as well as lithium batteries from computers and specialized ones from electric vehicles. The EPA estimated that the debris effort would be the largest lithium-ion battery cleanup in the world. There was no cost to the property owner for this phase. In less than thirty days, wildfire hazardous debris cleanup was completed at over 9,000 properties, or at 99 percent of the sites.

All household hazardous waste was taken to staging areas before being moved to other locations for storage, decisions that met with public opposition. Demonstrators protested against one sorting site at Will Rogers State Beach, and at another EPA site using a parking lot near the Malibu Civic Center as a temporary storage location. Residents attended public meetings to object to the selection of Lario Park as a waste site despite assurances from EPA officials that the hazardous waste would be properly packaged and transported. The site, owned by the ACE, is located in a working-class community of Black and Hispanic residents. Chants of "Find another place" and "Do the right thing" were heard at one town hall meeting, according to the *New York Times*. One real estate agent told the *Times*, "I understand they need to put the toxic waste somewhere. But it feels like they chose our city because they thought we wouldn't say anything." The protests and complaints epitomized the NIMBY concept of "Not In My Backyard" as one community after another sought to have the debris moved to someplace else (Selig et al., 2025).

In phase two, which began thirty-five days after the fires started, property owners had to inform ACE whether they wanted the agency to clear their land of any remaining debris from the wildfire. There was also an option that allowed property owners to pay out-of-pocket or use their own insurance to pay a specialized contractor rather than having the government remove debris from their site. ACE selected eighteen facilities for disposal; recyclable concrete, metal, and trees were sent to separation sites.

The issue of ocean contamination arose when scientists onboard a research vessel during the fires detected ash and waste on the water as far as 100 miles offshore, according to researchers with the University of California. Since some of the properties that burned during the wildfires were along the coastline, high tides and rain pulled debris into the ocean with potential impacts on both the water and marine organisms. State water officials conducted extensive testing for total and dissolved metals, volatile organic compounds, microplastics, and other toxic compounds. Beachgoers were advised to stay out of the water. Long term, there are concerns about whether or not toxic contaminants will enter the food chain, with further research necessary.

An additional issue that resulted from backlash from residents and officials was whether the soil would be tested after the cleanup crews removed the debris from the fires. The ACE and FEMA said that in addition to the rubble and ash from the fires they would remove the top 6 inches of topsoil from properties but would not order follow up soil testing to confirm whether a site was still contaminated after the process was completed. FEMA said that the cleanup strategy was based on scientific best practices and the agency's longstanding policies. FEMA further noted that local governments were free to conduct additional testing if they chose to do so. While some property owners considered hiring private contractors to test the soil, others were outraged that safety concerns were being ignored in favor of speed. In response, Los Angeles County conducted soil testing at both the Eaton and Palisades sites, with air and water sampling included in the post-fire assessment.

The Los Angeles County Department of Health found that even after the top layer of soil was removed by the Army Corps of Engineers, toxins such as lead were still present above state health screening levels. Independent studies conducted by the *Los Angeles Times*, with results released in May 2025, found hazardous levels of arsenic, lead, and mercury on some sites that exceeded California's residential safety standards by more than three times (Los Angeles Times Staff, 2025). These metals can cause neurological damage, cancer, and respiratory and kidney damage. The reports led attorneys for those whose homes were damaged or destroyed in the Eaton Fire in Altadena to blame Southern California Edison for exposing their clients to toxic substances (Petersen, 2025).

Early on, speed and convenience were considered essential to the process of moving forward. The governor issued an executive order in early January 2025 that suspended parts of California's 1970s-era Environmental Policy Act and the California Coastal Act, ensuring homeowners that property tax assessments would not be increased for those who rebuild in the communities affected by the fires. On the television program "Meet the Press" the governor noted,

"California leads the nation in environmental stewardship. I'm not going to give that up. But one thing I won't give into is delay. Delay is denial for people: lives, traditions, places torn apart, torn asunder" (Mehta et al., 2025). The resulting disaster rebuild waivers include a provision that exempts property owners from the usual permits required by the California Coastal Commission as long as the new structure is sited in the same location and is not more than 10 percent larger or taller than the one that was destroyed. The governor's action made clear that the state wanted homeowners to go back to their old neighborhoods, regardless of the wildfire risk, and that the process should be expedited.

Wildfire Politics and Policy

Many within the fire community, from scholars to firefighters, believe that there is a science behind the decision-making that drives how wildfires are fought. While that may be true, equally important are the political forces that come into play when decisions are made about agency or department budgets, personnel, priorities, or even the buildings where employees work. Wildfires are complex beasts whose behavior is sometimes unpredictable and deadly, and the politics affecting how they are managed is no different. Examples of these conflicts abound that illustrate the collision between fire science decisions and political decisions.

One example is exemplified by candidate, and then president, Donald Trump. Using social media, he posted on August 5, 2018: "Calif wildfires are being magnified & made so much worse by the bad environmental laws which aren't allowing massive amount of readily available water to be properly utilized. It is being diverted into the Pacific Ocean. Must also tree clear [*sic*] to stop fire spreading." The post, which would be repeated in similar language in his second term, was criticized by fact-checkers who noted that it was hot, dry weather conditions that caused recent fires to spread, not access to water. Another pointed out that the president was conflating two very different issues—the long conflict between farmers and environmentalists over how much of California's water resources should go to agriculture, rather than to lakes and rivers, which has nothing to do with firefighting.

Trump's feud with California's governor, Gavin Newsom, who he referred to as "Newscum," was center stage throughout his presidencies. On October 17, 2018, as *Wildfire Today* noted, Trump boasted at a Cabinet meeting that the state "better get its act together, cause California we're just not going to continue to pay the kind of money that we're paying because of fires that never should be to the extent [*sic*]," tweeting on November 10, "There is no reason for these massive, deadly and costly forest fires in California except that forest management is so poor. Billions of dollars are given each year, with so many lives lost, all because of gross mismanagement of the forests. Remedy now, or no more Fed payments" (Gabbert, 2020).

The political feud with the state continued during a September 2024 campaign stop by presidential candidate Trump at a time when 100,000 acres were burning in wildfires in the mountains outside Los Angeles. During a fundraiser at his golf club in southern California, Trump accused Governor Newsom of holding back water from a nonexistent "very large faucet" in Canada that had caused the state's drought. He threatened to withhold federal disaster response funding from California saying he would force Newsom to agree to send more water from northern parts of the state to the farmers in the southern area. The governor called the candidate's comments part of a political vendetta (Bassler, 2024).

There was more at stake in the dispute than just political scrimmaging and posturing. President Joe Biden declared major disaster status for Los Angeles County during the January 2025 fires while he was still in office, approving funding for costs such as cleaning up debris, hazard mitigation, and financial assistance for survivors. But the funding promise was only for 180 days, well into Trump's term as president. After that, only 75 percent of eligible emergency response costs would be reimbursed. When the funding account runs out of money, it requires Congress to take action to fund longer-term recovery for Los Angeles. By September 2025, the governor's request had not produced any additional assistance, with speculation that the feud between the president and the governor was not over.

President Trump had already used his authority to deny disaster funding in similar situations. He ordered the Office of Management and Budget (OMB) to delay billions of dollars in disaster aid to Puerto Rico in 2017 after Hurricane Maria. He had denied Newsom's request for a disaster declaration and wildfire aid in October 2020 when six fires burned across the state. Newsom joined Trump's staff in pointing out that many Republicans lived in the damaged areas, noting that Trump's decision mostly hurt counties that supported the president, and the decision to withhold aid was promptly reversed (Koseff, 2025b). He had blamed the wildfires on the governor's "gross incompetence and mismanagement" and revived his threats to withhold disaster aid (Frank and Waldman, 2025).

Although it is not entirely unusual for federal grant money to be conditioned on certain actions by the recipient, such as mitigation measures after a hurricane, the feud led to an unprecedented action. While the 2025 fires continued to burn, Speaker of the House Mike Johnson, who represents disaster-prone Louisiana, floated the idea of putting some kind of conditions on federal disaster assistance to California (Frank and Waldman, 2025). The president got more specific when he prepared to fly to California to survey the damage from the fires. "I just want voter ID as a start, and I want the water to be released," referring to his insistence since 2016 that he had lost the vote in California because non-citizens had participated in the presidential election, and his demand that water in northern California be released to the south. "After that, I will be the greatest president that California has ever seen" (Koseff, 2025a).

There is a political and media strategy called "flooding the zone" that has been a hallmark of Trump's second term. The phrase refers to releasing a flurry of statements, press releases, and inconsequential policy announcements designed to overwhelm critics by their sheer number. While the executive orders released from the White House in 2025 covered nearly every sector of the federal government, the ones that impacted wildfires took on special meaning as the fires continued to burn in California. On January 20, the announcement of a federal hiring freeze affected the onboarding of thousands of seasonal federal firefighters, including ones from agencies that were called on to battle wildfires. The executive order exempted positions related to public safety, but hiring officials in at least one agency, the Bureau of Land Management (BLM), said federal firefighters were not exempt. The BLM, along with the two other agencies most responsible for fuel management and wildfire management, the National Park Service and the US Forest Service, employ more than 15,000 career and temporary workers who also assist other jurisdictions in emergencies as part of the National Incident Management system. Officials in some field offices involved in annual hiring were told to withhold all offers related to fire positions, a recruitment process that is lengthy and competitive (Soboroff, 2025). Federal workers who had been working offsite since the Covid epidemic were told to return to their offices, which became problematic for some as building leases for some federal agencies were not renewed.

The president followed with another round of executive actions in February 2025, announcing the firing of tens of thousands of other federal workers, including 17,000 wildland fire staff employees. Of particular concern was the loss of 3,400 USFS workers in boots on the ground roles and others employed as radio operators, trail crews, scientists, and others with the agency in roles considered essential and supplemental to emergency response and mitigation who were dismissed. The firings included employees on probationary status that included both new hires and those moved or promoted into new jobs who had years of public service experience and training. Some had been given awards for exemplary performance. With orders sometimes released one day and rescinded the next, agencies were left in disarray as managers struggled with what to do in what the media was referring to as whiplash. Some employees accepted the federal government's proposed buyout order, while others heeded orders to look for employment in the private sector. One firefighter interviewed for *The Guardian* noted that wildfires were already burning hotter and faster and "are already beyond what humans can manage. Now we have fewer humans to manage them. Fire season will get worse" (Canon, 2025).

For wildland firefighters and support personnel, the administration's efforts to slash the federal workforce left communities facing the consequences at a time when fire season was now year round. The president of the California Fire Chiefs Association told *Stateline* "The public needs to know that they're at risk. If the public knew all of this, they would lose their minds." A union official who had served fifteen years as a firefighter with the USFS added, "There's going to be firefighters that die because of this, there will be communities that burn" (A. Brown, 2025). Meanwhile, the battle over whether the president had the authority to reduce agency staffing was mired in competing court decisions.

Aside from the immediate impact on personnel, the series of executive orders affected wildfire-related activities in other ways at different levels. For tribal nations, seasonal firefighters are of the utmost importance, paid for with federal funding that suddenly was in doubt. About 1,500 employees and contractors working with the Environmental Protection Agency to clean up wildfire debris in Southern California were told there was no certainty about how long they would be allowed to work. The non-profit organization ProPublica reported that in Montana, the federal grants used to assist the state's Department of Natural Resources and Conservation in helping communities become more resilient were cut off. Firefighters scheduled to attend a national leadership training program were told on short notice that it had been cancelled. Crews had already arrived for the 150,000-acre prescribed burn in Everglades National Park and Big Cypress Preserve that was cancelled at the last minute despite funding under the Bipartisan Infrastructure Law and Inflation Reduction Act (Olalde, 2025). The cuts to agencies such as the National Oceanic and Atmospheric Administration were also problematic since many states rely upon the National Weather Service and other NOAA programs for information needed to plan for prescribed burns, red flag warnings, and lightning storms.

Misinformation is often used in political warfare, and in the case of the administration's pronouncements about the Los Angeles fires, the errors of fact made national headlines. In an executive order released January 24, 2025, the US Army Corps of Engineers (not the US military, as the president claimed) released 2.5 billion gallons of water being held behind dams at Lake Kaweah east of Visalia, California, and Success Lake near the town of Porterville. The president, posting on social media, wrote, "Photo of beautiful water flow that I just opened in California," claiming that the release of water could have stopped or prevented the fires in Southern California. Fact-checkers noted otherwise. The water was released during the winter into areas like Tulare Basin, where it would evaporate, rather than helping Central Valley farmers who needed the water to irrigate their fields in the hot, dry Central Valley summer. By the time the president "opened the valve" as he put it, the California fires had already been contained. But more importantly, Southern California receives its water from the state's own complex water management system, not from the federal pipeline in the north. There is no giant pipe to funnel water from north to south, and several water district managers said the unplanned release posed a danger to downstream users and communities who had not been consulted. The *Washington Post* reported that the ACE colonel responsible for releasing the water from the two reservoirs wrote a memo four days after the release noting how federal officials rushed ahead with the plan to release the water despite objections from California's elected officials and some local farmers. In the memo, the colonel wrote that he knew the water could not be delivered to Southern California directly, but that it had been done to satisfy the president's executive order (Washington Post, 2025).

The nexus between wildfires and politics was not happening just at the federal level. California's legislators and voters also entered the fray during the 2024 election cycle with Proposition 4, a ballot measure that authorized $10 billion in bonds for water, wildfire prevention, and protecting communities and natural lands from climate change. About $1.5 billion of the funding would go to create fire breaks near communities using prescribed burns and grazing to improve forest health, home hardening, specialized fire equipment, and deployment of early detection and response systems. The measure passed with 59.8 percent of the vote, framed as California's response to climate change and climate resilience, with less attention paid to the small percentage of funding dealing with wildfires. Media coverage of that element in the proposal tended to focus on thinning forests and clearing vegetation—a totally different way of framing the issue.

After the election, state legislators and other officials had mixed feelings about where the money should actually be spent in what one observer called "Go broad or go deep?" The bond measure, passed just before the wildfires in January 2025, offered some flexibility. Some officials sought to have an incremental approach that would link smaller projects together to maximize effectiveness. The governor's approach, which would allocate $325 million to programs over five years, included trying to secure federal matching funds with the money. The state fire marshal said that while home hardening was important, so was managing forests.

California's legislature does not have a strong track record when it comes to wildfire prevention legislation. In 2025, lawmakers introduced about ninety measures related to wildfires, about half of which related to mitigation. Many of the bills had been introduced in previous years and had failed for reasons ranging from cost to conflicts with environmental regulations. In 2024, the governor vetoed two bills related to wildfire prevention, stating that they were too expensive. He also vetoed a bill that would have reduced the protection of endangered species so that it would be easier to clear vegetation. The bill was reintroduced in 2025 after the fires (Kamal, 2025).

Congress was not inattentive to the impact of the 2025 wildfires, especially representatives from the West. Dozens of bills were introduced to deal with every conceivable aspect of the issue, from ones dealing with wildland firefighter paychecks, fire-safe electrical corridors, and the creation of a Wildfire Intelligence Center to the bipartisan Fix Our Forests Act that called for the restoration of forest ecosystems, streamlining environmental reviews, more stakeholder partnerships, the building of fire resilient communities, and the acceleration of the removal of hazardous fuels. While their congressional sponsors attempted to move these somewhat lofty measures through the appropriate subcommittees, the president was thwarting their efforts by issuing frequent executive orders that did not require months or even years of legislative negotiation. While some actions involving cutbacks in personnel from agencies tasked with combating wildfires ended up in the judicial branch, others implemented a review of federal regulations and the revoking of Biden-era policies and initiatives. With the stroke of a pen, environmental legislation, grants and funding, research programs, and boots on the ground were eliminated in just minutes.

Wildfire Mitigation

The pronouncements about what leaders could have done to reduce the loss of life and property damage from the fires in Maui and California were delivered as fast and as furious as the flames that devastated their communities. Many characterized their situation as a perfect storm that combined hurricane-force winds, unseasonably dry conditions, and the onset of multiple fires at once that led to an inevitably destructive outcome. But there were also calls for investigations and reform, organizational change, and innovation to fight wildfires that can be totally different from those Smokey Bear warned about.

Myopia and externalities. Those two words were used by the authors of a 2021 working paper produced for the National Bureau of Economic Research to explain why, despite escalating disaster losses and predicted increases in weather-related catastrophes, communities have failed to adopt protective technologies and behaviors to respond to wildfires. The study identified several reasons why this is so, ranging from the fact that homeowners misperceive disaster risks and thus devalue the cost of investing in protective measures, to public spending on disaster preparedness that serves as a disincentive to private spending (Baylis, 2021).

Nowhere is the issue of wildfire in an urban setting better illustrated than in Mike Davis's 1998 book, *Ecology of Fear: Los Angeles and the Imagination of Disaster* in which he explores the history of catastrophic fires in Malibu. In the chapter, "The Case for Letting Malibu Burn," Davis questions why officials were averse to fire-risk zoning models, choosing instead to promote development and upper social succession with the combination of a federal disaster declaration and tax relief and low-interest loans for the residents. "Each new conflagration," he wrote, "would be punctually followed by reconstruction on a larger and more exclusive scale as land use regulations and sometimes even the fire code were relaxed to accommodate the fire 'victims'" (Davis, 1998).

Malibu is indicative of what Davis called the perverse fact that after every major California wildfire, homeowners and public officials conclude that they can keep building in fire-loving hills in the belief that if wildfire cannot be prevented, it can at least be tamed. After the 1993 firestorm, both Malibu (recently incorporated as a city) and Los Angeles County passed regulations requiring brush to be cleared around structures using fire-resistant roof materials. The concept of defensible space took over in place of the debate whether homes ought to be built in fire-prone areas at all.

Building Codes, Defensible Space, and Home Hardening

The city of Boston was the first to develop "building acts" designed to keep the city from burning down from a catastrophic fire. The ordinances, adopted during colonial times, were later expanded to include standards for plumbing and complete building plans. In 1866, the National Board of Fire Underwriters (NBFU) was created to find ways to reduce the amount of damage done by fires and the subsequent losses underwriters faced. The organization, followed by the Building Officials and Code Administrators in 1915, developed the first model building codes that made recommendations to state and local governments, but many decided to write their own regulations. Eventually officials decided that standardized building codes were needed to keep people safe. Today, these codes are the regulations that builders and architects follow when constructing or renovating a structure, covering safety, structural stability, and energy efficiency. Some codes have been developed at the state level, while others were created so that cities and counties could make their standards stricter than what the state had done. Codes were often influenced by disasters, such as the Chicago Fire of 1866, the Great Fire of Baltimore in 1904, and the 1906 San Francisco Earthquake and the subsequent fires that destroyed 80 percent of the city. Events like these led to major changes in building codes and construction methods and led to zoning ordinances that determined what could be built and where.

California led the way in adopting modern building codes for wildfire resistance after the Oakland Hills Fire in 1991 that led to $1.5 billion in property damage and the loss of twenty-five lives. The state legislature responded by requiring that vegetation near a house be maintained, and that special materials be used for roofs. In the 1992 Bates Bill, communities that had their own fire departments (called Local Responsibility Areas) were encouraged to have stronger building codes if the state's fire agency, CalFire, determined they were in Very High Fire Hazard Severity Zones based on maps the agency developed. In other areas of the state (State Responsibility Areas or SRAs), CalFire provides firefighting services and also develops building standards. Gradually, the requirements for roofing materials were upgraded to require that they be more ignition resistant, covering new construction or re-roofing projects. Building codes in California were revised again in 2008 to include, among other provisions, that exterior siding, vents, windows, doors, and decks be constructed of non-combustible materials. The vast majority of homes in the Los Angeles area were built before those codes went into effect, however.

In determining a structure's resilience to potential from a wildfire, technical experts in the fields of engineering and forestry use the term "home ignition zone" to describe the design of the home itself as well as an imagined area that extends 30 meters away from the structure (a distance of just under 100 feet). The goal is to eliminate all flammable materials within the zone, meaning both how the home is built and vegetation around the home. Another way of thinking about this area is the term "defensible space."

Only a handful of states have building codes specific to wildfires, including Nevada, Pennsylvania, Utah, and California. Wyoming has a statewide building code, but not all jurisdictions have adopted building codes of their own. Most states have been unsuccessful in attempting to require wildfire building codes. As National Public Radio reported in 2020, Oregon's Residential and Manufactured Structure Board began pushing for them in 2018 but faced pushback from the home construction industry. The group's members argued that wildfire building codes were unnecessary and that Oregonians should have the freedom to choose "where they want to live and the personal responsibility to construct their homes to work with that choice." The Oregon Home Builders Association testified that the proposed codes would add 5 percent to the cost of building a new home, potentially costing tens of thousands of dollars. But one of the local fire chiefs questioned those numbers, calculating the additional cost would be less than $1,700. The codes were eventually approved, but made optional, allowing cities and counties to choose whether to adopt them. In a twist of fate, 5,400 structures were destroyed in Oregon that same year in some of the state's most destructive wildfires. If the owners of those homes decide to rebuild, they will not be required to meet wildfire codes (Sommer, 2020).

Home hardening has become a common phrase in wildfire mitigation. In Colorado, two major fires in 2013 led the governor to convene a task force to examine the state's fire policy. The group's final report recommended the use of fire-safe materials as one of the most effective measures that homeowners could take, and called for a statewide model wildfire building code, citing encouraging results in counties that had already adopted one. As had been the case in Oregon, home builders' groups said that state regulation would be ineffective, and that local governments should determine their own codes and use outreach to homeowners. While some cities and counties decided to adopt wildfire building codes, some specified only roofing materials, while others decided not to adopt a code. After the 2021 Marshall Fire, which destroyed a thousand homes, the state legislature adopted a measure in 2023 that created a Wildfire Resilience Code

Board, charged with developing a statewide wildfire code. One of the key features of the Colorado code is the maps created with the assistance of the Colorado State Forest Service that determine an area's wildfire risk level. The maps have been met with criticism because the wildfire severity level is based on vegetation. Downtown Aspen, for instance, would get no wildfire risk rating at all because it lacks any vegetation, while others would be defined as low, medium, or high risk.

Vegetation management can include either a carrot or a stick approach to implement desired policies. Regulations usually require homeowners to space large bushes apart, clear their gutters of debris, and make sure tree limbs do not touch the exterior of a house. The responsibility for checking compliance usually falls to local fire inspectors, with a notice to a homeowner to take appropriate action. Without an inspection force (the stick) the carrot is more likely to include public education campaigns and voluntary guidelines, which are of less value unless an entire community gets involved. The most extreme measures, which call for the severe limitation or elimination of all vegetation next to a home, may be the next step in fire-prone areas, although the regulations would face strong pushback from homeowners who consider landscaping part of their home's aesthetic and price value.

Cost is one of the major considerations mentioned by homeowners considering retrofitting their house to make it wildfire resistant. In 2024, researchers with Headwaters Economics analyzed what steps, at what costs, it would take to retrofit a 2,000 square foot existing home in California in order to meet California's Building Code Chapter 7A, which deals with materials and construction methods of ignition construction. They found that it is possible to affordably reduce structure vulnerability to wildfire. This refers to steps such as removing flammable materials from decks, clearing out gutters, and removing vegetation and debris from the roof that have little to no cost involved. Minimal retrofits can cost as little as $2,000 to more than $100,000 if the highest level of protection is needed. Most homes do not need a full retrofit, however, and targeted replacement of specific components, such as replacing exterior vents with flame and ember-resistant vents, installing metal flashing at all deck-to-wall intersections, keeping gutters clean, and replacing bark mulch with non-combustible gravel go a long way to reduce risk (Barrett and Quarles, 2024).

One debate rages over whether insurance companies should provide financial incentives to homeowners who retrofit their homes in wildfire-prone areas. In California, legislation to do so calls for the retrofits to be cost-effective, while there are differences of opinion whether homeowner investments to improve structure survival and resilience are warranted. Should wildfire mitigation measure be voluntary or mandatory? Researchers are just now looking at whether there are spillover benefits of wildfire mitigation investments to those living on neighboring properties.

Infrastructure and Utility Hardening

The reports of dry fire hydrants in the midst of the January 2025 fires in Pacific Palisades and Altadena, California, were followed by lawsuits over water supply and delivery, and post-fire investigations to determine why a nearby reservoir was out of service due to repairs when the wildfires started. Residents filed suit claiming that the City of Los Angeles and the Los Angeles Department of Water and Power (LADWP) had designed a water supply system that had insufficient water pressure to fight an urban wildfire. In response, attorneys for the LADWP relied on a California Supreme Court decision dating back to 1911 to argue that water utilities could not be held liable for a failure to provide water to fight fires unless that had a specific contract to do so, which one of the attorneys suing the utility said "didn't make sense" (Saavedra, 2025).

Most experts agreed, however, that even fully functioning hydrants would not have been enough to battle fires of the magnitude that firefighters faced. One-quarter of the hydrants within the perimeter of the Palisades Fire were outdated models that dated from the 1940s with 2.5-inch outlets that the American Water Works Association says are no longer suitable for normal fire protection. Even modern 4-inch models might not have made a difference with a fire that expanded exponentially at the same rate as running a four-minute mile. The heat was so intense that it led to a cascade of infrastructure failures, with pipes and sewer systems damaged, power outages, and roadways ravaged. The group Santa Monica Architects for a Responsible Tomorrow wrote, "This crisis demands a fundamental reimagining of urban wildfire defense. As climate change brings longer, more intense fire seasons, Los Angeles must upgrade its infrastructure not just to current standards but to meet future challenges" (Santa Monica Architects for a Responsible Tomorrow, 2025).

The situation is not new. During Colorado's Marshall Fire in 2021, the city's hydrants experienced low to no pressure, as was the case in previous fires in Malibu. In the LA fires, one water resource specialist noted that there was "no water system in the world that would be prepared for an event like this one." The CEO for the Los Angeles Department of Water and Power said that water use peaked at four times the usual rate. "We are fighting a wildfire with urban water systems and that is really challenging. This is an unprecedented event" (de Puy Kamp et al., 2025).

These comments illustrate the types of infrastructure issues that have arisen as more intense fires erupt in urban areas and in the wildland-urban interface. In the case of the water supply, for instance, *High Country News* noted that today's urban water systems are designed to protect a few houses or an apartment complex from a fire, not block after block of an entire neighborhood, with everything burning at once. The National Fire Protection Association adds that the amount of water required during a fire is calculated based on a single structure, with additional amounts for fires spreading to adjacent structures, not for fully involved buildings or for approaching wildfires. That typically means the infrastructure system delivers 25 to 40 percent more water than needed (Mohr, 2025).

The cost of redesigning the water infrastructure—hydrants, pipes, tanks, and other improvements necessary to substantially increase the water flow for an intense urban fire similar to what Los Angeles has been experiencing would be incredibly expensive. Dr. Erik Porse, director of the California Institute of Water Resources, says that although the conditions surrounding the LA fires made fighting them nearly impossible, it is time to rethink some systems from an engineering perspective to determine how protection can be provided against fire events. But he cautions that improvements need to be factored into cost-benefit analyses before future events are made. "We need to always think about infrastructure that aligns without ability to finance it, and the true mitigation of risks those changes can offer us" (Sukel, 2025).

Utility hardening is not a recent development, but it does appear to be in the spotlight more often as a result of recent fires and litigation. Pacific Gas & Electric, for example, notes that its 10,000-mile undergrounding program is the largest effort in the United States to reduce wildfire risk by burying powerlines underground. The company also says it is using stronger poles and covered powerlines along with vegetation management near powerlines. In April 2025, Southern California Edison (SCE) announced its plans to bury more than 150 miles of powerlines in areas near Altadena and Malibu, a proposal that would require the approval of state regulators. As planned the cost would be between $860 million and $925 million, still less than the over $1 billion the utility company estimated it would cost to rebuild the infrastructure damaged or destroyed in the January 2025 wildfires. The strategy is designed to reduce the future risk of wildfires as well as building back a resilient, reliable grid for the company's customers, SCE noted.

Wildfire Planning and Preparedness

After the Camp Fire that destroyed Paradise, California, one of the key questions raised was why so many people died. University of Washington researcher Phillip Levin noted, "Fire is natural. But the disaster happens because people didn't know to leave, or couldn't leave. It didn't have to happen" (Wagner and Larson, 2019). In a 2019 study by the *Arizona Republic* and *USA Today* that looked at 5,000 communities across eleven states, an analysis of sixty-five risk factors found that 526 communities faced a wildfire risk factor greater than what had happened in Paradise. In the West, the analysis calculated the ratio of households to major exit roads; in Paradise, for instance, an estimated 1,818 households would be expected to use the six potential escape routes during a mass evacuation. But when the fire approached, demand was even greater because one of the identified exit roads on the town's evacuation map had the warning "Do Not Travel" and other routes were effectively cut off as residents tried to leave the area. The analysis also found that people who are older or disabled need more time to flee, a factor that may not always be considered, as was the case with the Tubbs Fire in Santa Rosa, California. Another important element of wildfire preparedness—phone notification systems—is becoming more common, but in Butte County, where the Camp Fire burned, the system had the capacity to broadcast messages but the county had never activated the system prior to the fire (Wagner and Larson, 2019).

One planning tool that has been well established in California for more than a decade is hazard maps designed to predict where a wildfire is likely to occur. In 2025, CalFire released the final round of computationally modeled local zones that cover areas where local fire departments are responsible for managing wildfire. The color-coded maps distinguish between hazard zones where the risk is "moderate," "high," and "very high," but do not cover areas in the state that are federally owned—about 45 percent of California—nor areas where CalFire is in charge. The maps are important because they determine the rules applicable to new construction. In those areas mapped as "high," for instance, there is a requirement that newly constructed homes must meet heightened fire-resistant building standards, and anyone who wants to sell their home must disclose the elevated fire risk to potential buyers. In "very high" hazard zones, the rules additionally require measures such as clearing dead trees, brush, and other flammable material, and new subdivisions must be designed with multiple evacuation routes, an adequate water supply infrastructure, and built-in fuel breaks in place. In theory, the hazard maps can help jurisdictions determine *where* wildfire mitigation efforts should be in place.

The 2025 maps indicate that about 3.7 million Californians live in "high" or "very high" hazard areas, about 3,626 square miles of land equal in size to the state of Delaware. Another way of thinking about how many people live in these areas is that about 1-in-10 Californians now live in places where rules dictate how and where they live, how leaders plan communities, and how housing and insurance markets function in the state's neighborhoods most likely to burn, according to an analysis by *CalMatters*. The two highest categories of hazard have increased by 168 percent since 2011 (Christopher, 2025). A 2023 study done by Resources for the Future found that homes in the "high" category, where sellers must disclose the level of fire risk, sold for 4.3 percent less on average than comparable homes nearby that were outside the hazard zone.

Sometimes wildfire preparedness falls into the "duh" category, such as enacting parking restrictions on red flag days, or limiting street parking at all times to allow emergency vehicles access. In one wildfire after another, from California to Hawaii, the graphic shots of charred vehicles stuck on roadways are a reminder of how people died as key roads become impassable, even for first responders. Even in communities where an escape or evacuation plan is in place, the sheer number of vehicles that hit the roads all at once can overwhelm thoughts of zones and warnings.

Another key planning tool lies in getting the warnings out soon enough. In some cases, those without access to cellphones or the internet may not be able to receive a wildfire warning message or alert. Some decide not to pay attention to the message to leave, even when simplified to the "Ready, Set, Go" format now being introduced. If a person decides to fight the fire themselves, or believes it is more important to protect their property than their life, even if they change their mind and decide to leave, it may be too late. Individuals who make last-minute decisions will strain the system and will become the most vulnerable because they end up at the back of a long line of others. In the Paradise Fire, where residents in the community's fourteen zones had been told to wait until directed to leave, the flames outran any warning that could be issued. Because there was no alarm siren to warn them, they assumed they were safe. Nearly all those who perished in the fire died in their homes.

Some communities have led the way in preparing for wildfires by forming a Fire Safe Council or becoming part of the Firewise USA network. The first category of organization is initiated at the state level, with local councils formed to implement projects such as fuel reduction and home hardening. California's program is one of the largest with as many as 300 communities participating. The National Fire Protection Association (NFPA) is the umbrella organization for the Firewise USA program, which works primarily with homeowner associations and individual communities. Of the 44.8 million homes and 70,000 communities estimated to be located within the wildland-urban interface in 2024, the NFPA said 2,283 were in good standing in the program. Community Wildfire Protection Plans (CWPP), in contrast, are typically comprehensive documents prepared by relevant stakeholders and presented in public meetings. Most CWPP have been developed in the West, with an estimated 1,000 currently in place to help participating entities identify wildfire risks, adapt to wildfire, and enhance resilience. An excerpt from one example of a CWPP from Fergus County, Montana, is found in Chapter 5. The plans are not legally required and often take a long time to prepare and get stakeholder participation and buy-in, but even starting one can make a difference. Neither the city nor the county of Los Angeles had a CWPP in place prior to the January 2025 wildfires.

What Else Does the Future Hold?

The analytics firm First Street has identified areas of California considered to be "climate abandonment zones" where communities will experience rising home insurance premiums, declining populations, and decreasing property values, averaging 6.2 percent through 2055. The most serious county on the list is Fresno County, with home value losses projected at 46 percent over the next thirty years, with home insurance rates skyrocketing 56 percent. While wildfires will continue to pose a major threat in California and some areas in the West, flooding is the largest natural hazard risk nationwide, costing $800 billion since 2000, representing two-thirds of natural disaster risk damages. Flooding now accounts for 90 percent of natural disasters in the United States, according to the Consumer Financial Protection Bureau. Climate change, the US Congressional Budget Office says, will increase flood damage by one-quarter to one-third by 2050, in inflation-adjusted dollars. Unlike the megafires that have been largely limited to a handful of states in the West, widespread flooding has occurred across large swaths of the United States, from coastal border states to parts of Appalachia. The good news? The First Street report has also identified some areas of the country that are considered climate-resilient, representing about 5 percent of census tracts clustered in the central United States. The top five in the report are Denver County, Colorado; Arapahoe County, Colorado; Dane County, Wisconsin; Douglas County, Nebraska; and Johnson County, Kansas.

Studies like these are important for several reasons as we attempt to predict future challenges. The information provides policymakers with political ammunition for choosing to support some projects over others, should they decide to pay attention to what the studies and projections show. The data may convince legislators representing California, for example, to seek more funding for future or potential disasters, such as flood control, rather than for imminent or recent disasters such as wildfire damage or mitigation. Officials from other states may seek preemptive funding, or they may ignore what researchers predict as the consequences of climate change and instead seek to protect their constituents from more readily identifiable harm.

Researchers are taking a second and even third look at fuel reduction strategies to determine if new information might affect how wildfire intensity might be managed. Using satellite remote sensing to study more than 700 reburn fires over the past fifty years throughout the western United States, a University of California, Davis, study has shown that an initial wildfire can have a moderating effect on future fires, called reburns, a finding that could determine future fuel treatments. By analyzing how an initial fire burned, for instance, forest managers could leverage lower-severity fires as a kind of initial fuel reduction treatment, they found. In a way, a low-severity fire could provide more fuel reduction coverage than mechanical thinning and burning.

University of Arizona researcher Donald Falk argues that forest managers need to do more mechanical thinning and prescribed burning by at least tenfold, citing the Four Forest Restoration Initiative as one example of what it would take. The Forest Service-led program, which started in 2010 and is known as 4FRI, has treated 720,000 acres as of January 2025. Some treatments involved commercial logging, noncommercial thinning, and burning. But even Falk notes that the government appears to be shifting priorities from prevention to firefighting. "Unfortunately," he said, "we have a pyro-industrial complex" (Loomis, 2025).

Technological advances are not just on the horizon—they are already being implemented. Many researchers and commercial firms are searching for ways to mitigate wildfires through public-private partnerships. In Nevada's Lake Tahoe

Basin, BurnBot, a remote-controlled, diesel-powered machine is being used by the Tahoe Fund to chew up 6-to-8-inch forest fuels. BurnBot can clear 22 acres in three days, compared to a hand crew that would take fifteen days to do the same work. The machine is designed to remove 75 percent of the forest fuels, leaving behind 25 percent—one desired forest prescription. The company that makes the device has others that can put the biomass from the BurnBot back into the soil and some that work on steep slopes. They have also developed an unmanned drone that can be used to start prescribed burns at the rate of up to 2,000 acres per day.

In Texas and Louisiana, where prescribed burns are commonly used, researchers found that by using Small Unmanned Aerial System (UAS) technologies to support prescribed fire activities using aerial ignition, managers were able to increase the number of hectares burned by 129 percent. A 2023 report comparing non-UAS burning from 2012 to 2018 and after the incorporation of UAS burning between 2019 and 2021 found that UAS burns were more efficient, enabling what they called previously unachievable efficiency when measured in terms of hectares burned each year. Calling the UAS a viable tool for accomplishing safe and more effective prescribed burn operations, the research shows promise for adding another strategy to the wildfire manager's toolbox (Lawrence et al., 2023).

Another technological advance, internet-connected sensors placed in strategic locations where wildfires are predicted or have occurred, measure temperature, humidity, and wind speed—factors that lead to high wildfire risk situations. The sensors, developed by Dr. Anthony Wexler of the University of California, Davis, among others, are about the size of a human palm and detect specific conditions that then trigger the deployment of an unmanned aerial vehicle (UAV) that looks for other indications of a fire. The UAVs, like the UAS, also have the advantage of not putting a human's life at risk when flying over dangerous terrain.

On a larger scale, the USFS and Department of the Interior signed an agreement in 2024 with the National Oceanic and Atmospheric Administration (NOAA) to use the Geostationary Operational Environmental Satellite R-series data to rapidly detect and report wildfire starts. The project uses the satellite's advanced remote sensing capabilities from 22,000 miles above the Earth to improve both speed and accuracy in pinpointing ignitions and progression in real time. This expands the agencies' ability to spot wildfires before they can be detected on the ground or reported to emergency operators, picking up the wildfire's exact location when smoke is reported. That information is key because it helps officials see government jurisdictional boundaries, fuels data, and incident perimeters. Monitoring smoke plumes helps officials direct firefighting efforts from the air, even when poor visibility hampers the deployment of airplanes and helicopters. Smoke plumes are also useful to public health officials who provide air quality information to the media and other agencies. In conjunction with NOAA, the National Weather Service (NWS) is using artificial intelligence (AI) to identify fires using the satellite data to send information to NWS channels to issue fire warnings to agencies. The system was first used in 2023 near Wichita Falls, Oklahoma, to alert incident commanders to move firefighting crews to a safer location just before a dangerous wind shift, and has subsequently been used in other wildfire situations (US Department of the Interior, 2024). These types of science-based programs came under fire under the second Trump administration, with some being de-funded while others faced heavy staff cuts that effectively ended their ability to function.

Technological advances aside, wildfire's challenges will persist unless and until policymakers agree that real dangers and risks exist, agree to staff agencies with the kinds of professionals sought after by other government agencies, make the hard decisions related to the wildland-urban interface, and put their political differences on the back burner. The Trump administration made it clear, however, that it planned to make even more far-reaching changes to the nation's wildland firefighting efforts by seeking to combine the nation's existing five agencies with firefighting jurisdiction into a single agency. The proposed Federal Wildland Fire Service would replace the thousands of staff in the Department of Agriculture's Forest Service and the agencies within the Department of the Interior in a major restructuring plan that would purportedly centralize and streamline firefighting. In a separate executive order, the president told the Secretary of Defense to prioritize the sale of any excess aircraft and parts to support wildfire mitigation and response, with the president calling for a quicker response to wildfires.

The proposal was presented to Congress in mid-2025 as part of the administration's budget bill after the devastating January 2025 wildfires in the Los Angeles area, although no figures were released that would indicate how much money the plan would cost or save. A similar proposal had been made in 2008 but failed to get sufficient political support after a report by the Congressional Research Service found significant drawbacks in the plan. Critics of the new proposal, including several former Chiefs of the US Forest Service, argued that the merger could potentially "increase the likelihood of more large catastrophic fires, putting more communities, firefighters, and resources at risk" (M. Brown, 2025). Timing of the proposal came after thousands of firefighters and wildfire support staff had been fired or left their jobs, although no exact figure on losses had been released by the White House. Perhaps worse yet, the administration's actions came as agencies were planning and hiring for the upcoming fire season. By the end of June 2025, the country had experienced more than 33,000 fires in twelve states that burned 1.7 million acres, more than double the number of fires reported over the same period in the prior ten years.

Further Reading

Associated Press. 2025. "California Governor Asks Congress for Nearly $40 Billion for Los Angeles Wildfire Relief." *NBC News*. February 23. https://www.nbcnews.com/news/us-news/california-governor-congress-40-billion-wildfires-rcna193360/.

Auchterlonie, David L., and Jeffrey A. Lehman. 2023. *Running Out of Time: Wildfires and Our Imperiled Forests*. Herndon, VA: Amplify Publishing.

Barrett, Kamiko, and Stephen L. Quarles. 2024. *Retrofitting a Home for Wildfire Resistance. Costs and Considerations*. Headwaters Economics. Spring. https://headwaterseconomics.org/wp-content/uploads/2024/06/wildfire_retrofit_report_R5.pdf.

Bassler, Hunter. 2024. "'Watch Our State Burn': Trump's Threat to Withhold California's Wildfire Aid Angers State's Firefighting Force." *Wildfire Today*. September 15. https://wildfiretoday.com/2024/09/15/trumps-threats-to-block-california-wildfire-aid-if-elected-president-firefighter-reaction/.

Baylis, Patrick W. 2021. "Mandated vs. Voluntary Adaptation to Natural Disasters: The Case of US Wildfires." *National Bureau of Economic Research. Working Paper 29621*. https://www.nber.org/papers/w29621/.

Brown, Alex. 2025. "Trump's Cuts to Federal Wildfire Crews Could Have 'Scary' Consequences." *Stateline*. February 18. https://stateline.org/2025/02/18/trumps-cuts-to-federal-wildfire-crews-could-have-scary-consequences/.

Brown, Matthew. 2025. "Trump Plans to Merge Wildland Firefighting Efforts into One Agency, but Ex-Officials Warn of Chaos." Associated Press. May 20. https://apnews.com/article/wildfires-trump-fire-service-climate-change-85a0d6bfb15677c01a4b5fbe913eef85/.

California FAIR Plan Association. 2025. "Statistics and Data." https://www.cfpnet.com/key-statistics-and-data/.

Canon, Gabrielle. 2025. "Mass Firings Hamstring Federal Land Agencies and Wildfire Response." *The Guardian*. February 21. https://www.theguardian.com/us-news/2025/feb/21/trump/federal-land-wildfire-response/.

Christopher, Ben. 2025. "New Fire Maps Put Nearly 4 Million Californians in Hazardous Zones. What Does That Mean for the People Who Live There?" *CalMatters*. March 24. https://calmaters.org/housing/2025/03/calfire-maps-hazard-california/.

Davis, Mike. 1998. *Ecology of Fear: Los Angeles and the Imagination of Disaster*. New York: Vintage.

De Puy Kamp, Majlie, Curt Devine, Casey Tolan, Blake Ellis, Melanie Hicken, Rob Kuznia, Scot Glover, Yahya Abou-Ghazala, Audrey Ash, and Neil Black. 2025. "No 'Water System in the World' Could Have Handled the LA Fires. How the Region Could Have Minimized the Damage." *CNN*. January 10. https://www.cnn.com/2025/01/10/us/california-la-fires-emergency-prep-invs/index.html.

Dwyer, Mimi. 2025. "After Losing Homes and Businesses, LA Wildfire Victims Face a Hurdle to Rebuilding: Trump's Tariffs." *The Guardian*. April 10. https://www.theguardian.com/us-news/2025/apr/10/la-wildfire-cictimes-trump-tariffs/.

Fannin, Blair. 2024. "Preliminary Texas Panhandle Wildfire Agriculture Losses Top $123 Million, Costliest in State History." *Agrilife Today*. May 6. https://agrilifetoday.tamu.edu/2024/05/06/preliminary-texas-panhandle-wildfire-agriculture-losses-top-123-million-costliest-in-state-history/.

Frank, Thomas, and Scott Waldman. 2025. "Biden Approved LA Wildfire Aid. But Trump Will Control Spending." *E & E News*. January 14. https://www.eenews.net/articles/biden-approved-la-wildfire-aid-but-trump-will-control-spending/.

Fuller, Sarah. 2024. "Texas Wildlife Feel the Impact of State's Largest Wildfire." *Agrilife Today*. March 14. https://agrilifetoday.tamu.edu/2024/03/14/texas-wildlife-feel-the-impact-of-states-largest-wildfire/.

Gabbert, Bill. 2020. "Trump Administration Reverses Decision to Deny California's Request for Fire Disaster Assistance." *Wildfire Today*. October 16. https://wildfiretoday.com/2020/10/16/trump-administration-rejects-californias-request-for-fire-disaster-assistance/.

Gatsby Investment. 2025. "How Long Will It Take to Rebuild after the LA Wildfires?" https://www.gatsbyinvestment.com/blog/rebuilding-LA-after-the-wildfires/.

Gray, Audrey, and Andrew Robinson. 2025. "Smoke and Ash Made More Toxic by the Contents of Burning Homes Threaten Residents of LA and Beyond." *Inside Climate News*. January 17. https://insideclimatenews.org/news/17012025/toxic/smoke/threaten/los/angeles/residents/.

Hanson, Chad. 2021. *Smokescreen: Debunking Wildfire Myths to Save Our Forests and Our Climate*. Lexington: University Press of Kentucky.

Institute for Applied Economics. 2025. *Impact of 2025 Los Angeles Wildfires and Comparative Study*. February. https://laedc.org/wpcms/wp-content/uploads/2025/02/LAEDC-2025-LA-Wildfires-Study.pdf.

Johnson, Lizzie. 2021. *Paradise: One Town's Struggle to Survive an American Wildfire.* New York: Crown.

Kamal, Sameea. 2025. "Why Bills to Help Prevent California Fires Fail." *CalMatters.* February 27. https://calmatters.org/digital-democracy/2025/02/california-wildfire-prevention/.

Kerbeck, Robert. 2019. *Malibu Burning: The Real Story behind LA's Most Devastating Wildfire.* Rockford, IL: MWC Press.

Koseff, Alexei. 2025a. "Trump and Newsome Embrace in Fire-Ravaged LA, but Now the President Wants to Tie Federal Aid to Voter ID." *CalMatters.* January 24. https://calmatters.org/politics/2025/01/trump-la-fires-newsom/.

Koseff, Alexei. 2025b. "How Gavin Newsom Is Balancing the Fight against Trump with Los Angeles Fire Relief." *CalMatters.* April 24. https://calmatters.org/politics/2025/04/trump-newsom-california-wildfires/.

Lawrence, Brett L., Kevin Mundorff, and Eric Keith. 2023. "The Impact of UAS Aerial Ignition on Prescribed Fire: A Case Study in Multiple Ecoregions of Texas and Louisiana." *Fire Ecology* 19: 11. https://fireecology.springeropen.com/articles/10.1186/s42408-023-00170-x#Abs1/.

Li, Zhiyun, and William Yu. 2025. "Economic Impact of the Los Angeles Wildfires." *UCLA Anderson School of Management.* March 3. https://www.anderson.ucla.edu/about/centers/ucla-anderson-forecast/economic-impact-los-angeles-wildfires#:text=.

Loomis, Brandon. 2025. "Fight Fire with Fire? Thinning Forests Is a Start, but Experts Say It's Time to Burn More." *Arizona Republic.* May 11. https://www.azcentral.com/story/news/local/arizona-environment/2025/05/11/thinning-arizona-forests-helps-but-experts-say-controlled-burns-are-critical/.

Los Angeles Times Staff. 2025. "Uncovering the Toxic Soil Lurking in L.A.'s Burn Zones." *Los Angeles Times.* May 4. https://www.latimes.com/environment/story/2025-05-04/treated-like-dirt-soil-testing-project/.

Machado, Jessica. 2024. "Over a Year after Lahaina Fires, Native Hawaiian Homeowners Hit Another Breaking Point." NBC News. November 30. https://www.nbcnews.com/news/asian-america/year/lahaina-fires-native-hawaiian-homeowners-hit-another-breaking-poi-rcna180863/.

Mehta, Seema, Liam Dillon, and Rosanna Xia. 2025. "Newsom Suspends Landmark Environmental Laws to Ease Rebuilding in Wildfire Zones." *Los Angeles Times.* January 12. https://www.latimes.com/california/story/2025-01-12/newsom-suspends-landmark-environmental-regulations-palisades-altadena-fires/.

Mendoza, Candyd. 2025. "California Homeowners Brace for Insurance Rate Hikes after $1bn FAIR Plan Bailout." *Mortgage Professional.* February 14. https://www.mpamag.com/us/news/general/California-homeowners-brace-for-insurance-rate-hikes-after-1bn-fair-plan-bailout/524824/.

Meyersohn, Nathaniel, and Anna Bahney. 2024. "The Home Insurance Market Is Crumbling. These Owners Are Paying the Price." *CNN*. March 29. https://www.cnn.com/2024/03/29/economy/home-insurance-prices-climate-change/index.html.

Mohr, Kylie. 2024. "When a Utility Sparks a Wildfire, Who Pays?" *High Country News*. July 1. https://www.hcn.org/issues/56-7/when-a-utility-sparks-a-wildfire-who-pays/.

Mohr, Kylie. 2025. "Water Systems Aren't Designed for Unlimited Demands during Wildland-Urban Interface Fires." *High Country News*. January 19. https://www.hcn.org/articles/wildfires-are-too-much-for-municipal-water-systems-in-los-angeles-firefighters-tried-anyway/.

National Centers for Environmental Information. "National Oceanic and Atmospheric Administration." 2025. U.S. Billion-Dollar Weather and Climate Disasters. https://www.ncei.noaa.gov/access/billions/, DOI: 10.25921/stkw-7w73.

Oh, Sangmin S., Ishita Sen, and Ana-Maria Tenekedjieva. 2024. "Pricing of Climate Risk Insurance: Regulation and Cross-Subsidies." Harvard Business School Working Paper 24–077. https://www.hbs.edu/ris/Publication%20Files/24-077_b2916af5-696e-4a2b-aide-936aa2c876a5.pdf.

Olalde, Mark. 2025. "How Trump's Federal Funding and Hiring Freezes Are Leaving America Vulnerable to Catastrophic Wildfire." *ProPublica*. February 14. https://www.propublica.org/article/trump-funding-freeze-wildfire-season/.

Petersen, Melody. 2025. "Edison Hit with Lawsuit Saying Eaton Fire Exposed People to Toxic Substances." *Los Angeles Times*. May 8. https://www.latimes.com/environment/story/2025-05-08/edison-sued-for-dangerous-lead-levels-left-after-eaton-fire/.

Pope, Kristen. 2024. "California's New Megafires Are Taking a Toll on Wildlife Habitat, Researchers Find." *Yale Climate Connections*. March 18. https://yaleclimateconnections.org/2024/03/californias-new-megafires-are-taking-a-toll-on-wildlife-habitat-researchers-find/.

Redfin. 2025. "Pacific Palisades, CA Housing Market." May. https://www.redfin.com/neighborhood/64066/CA/Los-Angeles/Pacific-Palisades/housing-market/.

Reyes-Velarde, Alajandra, and Jeanne Kuang. 2025. "'It All Ended in a Second: Thousands of Low-Income and Immigrant Workers Lost Jobs in LA Fires." *CalMatters*. January 17. https://calmatters.org/environment/wildfires/2025/01/la-fires-workers-lost-jobs/.

Saavedra, Tony. 2025. "LADWP Argues It Can't Be Sued for Lack of Water to Fight Palisades Fire." *Whittier Daily News*. April 15. https://www.whittierdailynews.com/2025/04/14/ladwp-argues-it-cant-be-sued-for-lack-of-water-to-fight-palisades-fire/.

Santa Monica Architects for a Responsible Tomorrow. 2025. "Opinion: The Water Crisis Behind LA's Fire Disaster: A Legacy of Outdated Infrastructure." *Santa Monica Mirror*. February. https://smmirror.com/2025/02/sm-a-r-t-column-the-water-crisis-behind-las-fire-disaster-a-legacy-of-outdated-infrastructure/.

Schlichter, Sarah. 2025. "The Average Home Insurance Cost in the US for 2025." *Nerd Wallet.* January 2. https://www.nerdwallet.com/article/insurance/average-homeowners-insurance-cost/.

Scripps News Staff. 2024. "Wildfire Smoke Linked to Increased Risk of Dementia Compared to Other Pollutants." *KXLF.* July 29. https://www.kxlf.com/science-and-tech/climate-change/wildfire-smoke-linked-to-increased-risk-of-dementia-compared-to-other-pollutants/.

Selig, Kate, Jesus Jimenez, and Mimi Dwyer. 2025. "The New NIMBY Battle over the Waste from the LA Fires." *New York Times.* February 11. https://www.nytimes.com/2025/02/11/us/wildfires-toxic-waste-sites/.

Soboroff, Jacob. 2025. "Trump's Freeze Stalls Federal Firefighter Hiring." *NBC News.* February 6. https://www.nbcnews.com/news/-us-news/trumps-freeze-stalls-federal-firefighter-hiring-rcna191001/.

Sommer, Lauren. 2020. "Rebuilding after a Wildfire? Most States Don't Require Fire-Resistant Materials." National Public Radio. November 25. https://npr.org/2020/11/25/936685629/rebuilding-after-a-wildfire-most-states-dont-require-fire-resistant-materials/.

Sukel, Kayt. 2025. "LA Wildfires Magnify Need to Rethink Infrastructure." *Civil Engineering Source.* January 28. https://www.asce.org/publications-and-news/civil-engineering-source/article/2025/01/28/la-wildfires-magnify-need-to-rethink-infrastructure/.

Terrill, Marshall. 2025. "Measuring the Supply Chain Impact of the LA Fires." *ASU News.* January 21. https://news.asu.edu/20250121-business-and-entrepreneurship-measuring-supply-chain-impacts-la-fires/.

Tidmarsh, Kevin. 2024. "Oh Nuts! Wildfire Smoke Harms a Crucial Part of California's Economy, Study Says." *LAist.* October 6. https://laist.com/mews/climate-environment/wildfire-smoke-harms-a-crucial-part-of-californias-economy-study/.

Udasin, Sharon. 2024. "Wildfire Smoke Has Covered Up to 70 Percent of California in Recent Years, Affecting Land and Water: Study." *The Hill.* May 22. https://thehill.com/policy/energy-environment/4677656-california-wildfire-smoke-covered-up-to-70-percent-land-water-impacts-study/.

University of California, Los Angeles. 2024. "The Death Toll from Wildfire Smoke." *Press Release.* June 7. https://newsroom.ucla.edu/releases/the-death-toll-from-wildfire-smoke/.

US Department of the Interior. 2024. "Biden-Harris Administration Advances Early Wildfire Detection by Expanding Use of Satellites as Part of Investing in America Agenda." *Press Release.* July 23. https://www.doi.gov/pressreleases/biden-harris-administration-advances-early-wildfire-detection-expanding-use/.

Waggoner, John. 2023. "Why Homeowners Insurance Rates Are Going through the Roof." *AARP.* January 30. https://www.aarp.org/money/budgeting-saving/info-2023/homeowners-insurance-prices-soar.html.

Wagner, Dennis, and Pamela Ren Larson. 2019. "Where Will the West's Next Deadly Wildfire Strike? The Risks Are Everywhere." *Arizona Republic*. July 22. https://www.azcentral.cm/in-depth/news/local/arizona-wildfires/2019/07/22/wildfire-risks-more-than-500-spots-have-greater-hazard-than-paradise/.

Washington Post. 2025. "Army Corps Knew Trump Order Would Waste California Water, Memo Shows." *Washington Post*. March 7. https://www.washingtonpost.com/climate-environment/2025/03/07/trump-water-release-california-fires/.

3

Perspectives

Understanding wildfires requires a background in multiple disciplines, each with its own terminology, used by different types of stakeholders, in complex situations, and with sometimes conflicting biases. This chapter exemplifies those factors through the words of six writers whose experiences and knowledge provide different perspectives on their subjects. The first contributor, Cathy Barta, is a resident of a small Montana community who led the development of the community wildfire protection plan excerpted in Chapter 5. Tim Brown of Reno's Desert Research Institute seeks to find ways to apply what researchers have learned about wildfires into practical use and application. A new area of field study is seen through the eyes of Julie Dinasquet and Douglas S. Hamilton and their research on the impact of wildfires on oceans. A boots-on-the-ground perspective is offered by Dirch Foreman, an Arizona firefighter who has held many positions with fire agencies and has been on the front lines of wildfires in the wildland urban interface (WUI). Lastly, consultant Jeff Shelton pairs his experiences as a fire captain in Southern California with his ability to pare down the fire problem to what he terms a "hot mess" requiring more leadership and less bureaucracy.

Firefighters monitor a prescribed fire in Montana's Helena-Lewis and Clark National Forest, Fall 2023. Prescribed burns are a management tool used to reduce flammable materials that might otherwise result in wildfire, but the practice can be risky when sufficient planning has not taken place or when conditions such as weather change dramatically without notice (USDA Forest Service).

Building Resiliency with a Community Wildfire Protection Plan

Cathy Barta

In January 2025, Malibu, California, once again made headline news with wildfires burning homes and taking lives. I used to live in southern California, so I watched the areas being burned with interest. As a retired federal woodland firefighter, I am intrigued by the discussion that followed to try to "fix" the fire problem. I wish people would understand that we can't "fix" this problem. Fire is as impossible to fix as it is to stop an earthquake, tornado, or hurricane. Fire is a normal ecological process that North America has evolved with over thousands of years. Granted, it is not okay when people intentionally start fires. But the natural ones that originate from lightning will continue to happen and we need to adapt and learn to structure our neighborhoods to live with fire. We have chosen to live in a place where fire is a regular occurrence. We need to design and build our communities to be more resilient to this reality.

The current problem, as I see it, is that land developers have subdivided areas with small lot sizes to maximize profits with land sales. Cities and towns adopt building regulations which allow smaller setback requirements to allow homeowners to maximize the buildable area of their vacant lots. Most states do not regulate construction materials to be fire resilient, so the most cost-effective source, which typically is wood, becomes the primary building material. Homeowners enjoy sunlight and rooms with a view, so large windows are frequently incorporated into the housing design. Window coverings are common with high-density neighborhoods for privacy, and these are generally lightweight, fire-vulnerable materials. The temperate climate of southern California has drawn large numbers of people to live there. Unchecked development continues to tax existing water supplies. The carrying capacity of the area was exceeded decades ago, yet people still come; money still flows, and more and more houses continue to go up.

The media feeds a false sense of hope about a fix that fuels the politics of wildfire. But by reporting about the need for investment in the organizations that fight fire, we can help solve this problem and protect neighborhoods and lives. A few more fire stations, more hydrants in the neighborhoods, more fire fighters for staffing, more fire engines, and more retardant planes and helicopters will help solve the problem. With enough of a financial investment, we should be able to solve it. But it never does and never has. During dry periods and high wind events, there just isn't enough of anything out there to stop fire, no matter how much money you throw at it. Just giving the government(s) more and more money will not change the outcome during weather conditions like we are seeing now in southern California. And we will see these conditions again. And not just in southern California.

So, what is the solution? I believe the solution is not to think about trying to stop it, but to learn to live with it. Building our communities to be more resilient to fire can make a difference and give neighborhoods a chance. But what does that look like? As someone who recently co-authored a comprehensive Community Wildfire Protection Plan (commonly referred to as a CWPP), I believe there are immediate steps that individuals can take. Do you know if your city or county has prepared one of these documents, or how you can find it? If your community already has one, take a look at it and see if it makes sense and if it is still relevant to your community. Is it up-to-date? The plans are often prepared by a non-local company utilizing templates with generic information that may or may not apply to your area. They are required to be updated every five years and must involve the public in the process. If this is not the case, then make it so.

If your community does not already have a CWPP, start taking the steps to get one started. Invite everyone in your community, and be sure to include local government officials for both the city and county, invite other stakeholders, such as non-profits like the American Red Cross. Be sure to include the utility companies, such as water, power, and natural gas, to understand their service protocols during fires, public works for water and sewer considerations, mobile and wireless companies to ensure there is a plan for disruption of service and its impacts. Invite the school districts that may have bus service available to assist with evacuations, as well as local council on aging services to understand the impacts to non-ambulatory seniors to be able to develop a plan for their evacuation when needed. If large animals are present in the community, develop an evacuation plan with early trigger points, so the process doesn't impede the emergency services response to the fire. In mountainous terrain with limited access roads, consider prioritizing building a safe area of refuge, also known as a safety zone, when evacuation no longer is possible. Get involved in your community. It will be the best investment for the future of your community.

One component to think about that might help make a community more resilient is increasing the spacing between structures. Photos of the recent Malibu fires showed areas where the vegetation in the neighborhoods survived the fire, but the houses did not. The houses themselves became the fuel source because they were so close together. When it is not practical to increase the spacing, adopt building regulations that require fire-resistant material be used for all new additions and renovations, without exception. Develop options for protecting the interiors of housing against heat transfer through windows. Prioritize the protection of watersheds and water resources. Identify and communicate high hazard areas, and develop protection plans for them. People need to learn how to invest in and protect their property like no help is coming, because often there just aren't enough resources to go to every single house to help when it is needed most.

I hope the Malibu area will get serious about their CWPP. They will have more large fires, and if they don't do something different, they can expect similar results. In 2018 the *Los Angeles Times* published an article about the fire history of the Malibu area, and stated that over the last ninety years, there have been thirty major fires re-burning the same area. The headlines of 2025 come as no surprise. Hopefully, with community effort, they will read differently.

Further Reading

Fergus County, Montana. 2024. *Community Wildfire Protection Plan*. July. Fergusmt.gov/images/disaster/emergency/2024.07.31-final-cwpp-draft.pdf.

Catha Barta is the author of the Fergus County Community Wildfire Protection Plan.

The Interface between Wildland Fire Science and Decision Making

Tim Brown

The roots of fire science in the United States began in California in the early 1900s. Over the decades fire science became a formalized discipline, especially after the Second World War as research in general accelerated and surplus equipment from the war was utilized in fire suppression. For many decades the dominant research was related to fire suppression. The 1960s and 1970s brought changes to forest management philosophy beyond suppression, particularly with the establishment of wilderness areas and developing policies regarding restoring fire to Western landscapes after decades of fire exclusion. Following a research lull in the 1980s largely due to funding shortages, the 1995 Federal Wildland Fire Policy and the National Fire Plan in 2000 provided stimulation for wildland fire research. These policies opened the door for inclusion of social science fire research in what had been a physical science-dominated endeavor. In 1998 the Joint Fire Science Program was established and by around 2010 had developed the Fire Science Exchange Network, currently consisting of fifteen consortiums across the United States with a goal "to accelerate the awareness, understanding, adoption, and implementation of readily available wildland fire science information."

What is clear from the over 100 years of fire research evolution is the continued need to translate science information to guide fire management decision and policy outcomes. Fire is a very complex coupled socioecological system. Taught in wildland fire courses is the infamous fire behavior triangle that comprises fuel, weather, and topography. All three of these physical components are made up of numerous metrics deriving from measurements and models. The combination of these factors directly relates to fireline intensity, spotting, fire spread rate, flame height, and duration. But outside this triangle are factors that influence wildfire outcomes and actions, all having significant human elements. Start with the ignition source—this is either naturally caused (e.g., lightning) or human caused (e.g., trash and debris burning, power lines, arson, campfire), which makes up the largest proportion of fire starts. Fuel and fire management are key aspects of hazard reduction, prevention, and suppression. Fire management is ultimately mitigation focused on economic, environmental, and societal impacts. Cultures have always been connected to fire. One way to describe today's large and destructive fires is the confluence of climate, fuels, and people—society has responsibilities for all three factors. Thus, while any aspect of physical fire can be researched on its own, and there are at times good reasons to do this, the larger context that fire is a physical environment-societal connection should not be forgotten. For any information generated in a scientific structure, there is potential for a decision-making benefit. Sometimes the outset aim is to inform a decision or policy, other times it is scientific curiosity that drives the research.

How can science get adopted by management? There are three recognized fundamental types of research—basic, applied, and user-inspired. An example of basic fire science is modeling fire physics, that is, understanding the physical attributes of fire in a manner that can be described and predicted quantitatively in mathematical equations. Applied fire science might be either physical or social, such as a statistical analysis of weather conditions in relation to fire danger, or analysis of human behavior given some type of fire-related warning, or a combination of both, analyzing the requirements of fire staffing resources given the potential for numerous and large fires. While basic and applied research refers to fundamental and applied knowledge, respectively, these require a minimal role of end users (or users, defined as a practitioner utilizing the results) in the research process. Basic and applied approaches have a very important place in research, but it is often not obvious how the scientific output can or will be incorporated into decision-making, even though the result appears to be a valuable real-world contribution. Many conversations have taken place asking how science results can effectively be utilized, which is ultimately asking what kind of process is needed to translate science into an action. A common phrase for this process in many research disciplines is research-to-operations. It might also be referred to as translational science, or specifically here, translational fire science. This requires a focus on decision process as much as the tool or data development to help inform a decision.

A third type of research, user-inspired, more directly incorporates stages of decision-making, including developing and weighing alternatives, formulating decisions, and implementing decisions. This is a codevelopment process based on greater involvement and engagement with users throughout the entirety of the project. How might this work in practice?

Example 1. A user need has been identified to standardize the methodology for a National Weather Service (NWS) fire weather warning. Physical science research is undertaken that develops a quantitative methodology highlighting key factors to be used for informing the potential issuance of a warning. The research effort even achieves a successful master's degree for a graduate student. The output by itself, while interesting, will not be effectively used without direct involvement by NWS and fire management personnel from the beginning. The former (forecasters) being the providers of the product and the latter (managers) users of the product. For this approach to be effective, there must be integration of the physical science with the social dimensions to deliberately serve the need of the forecasters and decision makers. In this example, the project inception began by holding joint workshops to identify what the current product issues were, what outputs were needed, and what outcomes were desired. The physical science component represents less than half of the effort because once the tool/information is developed, it must go through an iteration process of testing before becoming operationally adopted, which then also includes product education and training for both the providers and the users.

Example 2. It is determined that a seasonal (three months combined) wildfire potential outlook might be beneficial for longer-term agency planning. A series of workshops are held comprising climate scientists and operational fire agency meteorologists that have direct connections to fire managers. These meteorologists have a key understanding of how the product might be used. Instead of having a "simple" seasonal forecast of just number of fires or area burned, this user knowledge led to a product highlighting the potential for agencies needing additional management resources in geographic locations across the United States, a wholly new concept and product. The workshop partners committed themselves to sustained interactions, equality in partnership, and clear partnership responsibilities. The product was tested over a few fire seasons, then operationalized. The success of the product was tested via social science methodology. Through a process of analyzing the meteorologist and fire management social network it was realized that these meteorologists are information nodes. That is, they are trusted and valued sources for information that will be utilized in a decision process, effectively the concept of "diffusion of innovation." From a researcher perspective, it is not necessary to bring every potential user on board and instead co-producing the product with key individuals who are respected in their network and have the desire to advance and promote the product. This will yield successful outcomes.

Despite efforts to work in collaboration, the phrase "working in silos" is commonly heard. In the context here it refers to researchers not realizing what other researchers are doing. Some latitude can be given for this because limited resources in an entrepreneurial environment, which is effectively how much of science is supported, can encourage collaboration and but can also exacerbate the silo effect through lack of peer communication. Research silos are also formed by not engaging practitioners. These individuals have substantial knowledge from their "boots on the ground" experiences. Indigenous knowledge is a prime example of substantial experience not being fully accepted in modern society. It is not uncommon to find a scientific paper that highlights a result as new, though it was already known in practice. This does not mean the paper has no value as there is high relevance to documenting information, confirming or refuting knowledge, and allowing others to build off that result. This is the scientific method.

But it does highlight a common gap in science inquiry and the use of the produced information in a decision process. This gap is not necessarily about the use of information by an individual but also refers more broadly to adoption of the innovation by an institutional culture that creates policies based on science. If the researcher goal is to develop knowledge for decisions and policy, then the likelihood of success requires communication and collaboration with the practitioner providing for a scientist-stakeholder partnership to achieve desired outcomes. New fire science such as remote sensing and artificial intelligence is rapidly emerging that could highly benefit management actions, but at the same time risk providing an overload of information and exceeding the capacity of the user. Especially given these technologies, the interface between fire science and decision-making should be even more thoughtfully considered.

Dr. Tim Brown conducts user-inspired research and applications development at the Desert Research Institute (DRI) in Reno, Nevada, and is Director of the Western Regional Climate Center.

Ocean on Fire: How Wildfires Both Nourish and Threaten Marine Life

Julie Dinasquet and Douglas S. Hamilton

When most people picture a wildfire, they imagine flames consuming forests or racing across savannah grasslands; scenes rooted firmly on land. Rarely do images of ocean waves or marine life come to mind. Although it might sound surprising at first to think of the ocean in connection to a wildfire, over the last decade scientists have begun to uncover the unexpected links between wildfire and oceans, revealing how wildfire is a key Earth System force that connects all the classical elements: fire, earth, air, and water. Current understanding of the impact of wildfires on the ocean has largely emerged from studies involving forest and grassland fires. These studies show that, under certain conditions, atmospheric nutrient inputs from fire can enhance ocean life. The connection between terrestrial wildfires and marine ecosystems arises through combustion-produced debris entering the ocean through multiple pathways, with distinct timings and mechanisms.

The most immediate impacts of wildfires on ocean ecosystems can stem from the atmospheric transport and deposition of nutrient-rich particles in smoke plumes. These aerosolized particles travel vast distances with the prevailing winds, settling on the ocean surface up to thousands of miles away from the fire, influencing both coastal and remote open ocean waters. Research has shown that fire smoke is an efficient vector for the transport and redistribution of essential nutrients such as nitrogen and phosphorus, and also trace elements such as iron. Those same nutrients that help a forest grow on land are also necessary for the ocean's microscopic plants: the phytoplankton. As the vegetation is consumed by the fire, the nutrients that once formed the building blocks for plant life on land are released in the combustion process, being lifted into the atmosphere in smoke, and transported to all corners of the globe. When they land on an ocean patch with little access to nutrients, these wildfire-derived nutrients can fertilize production in the ocean: bringing new life from death.

Delayed and chronic impacts of wildfires on ocean ecosystems can stem from post-fire erosion and coastal runoff carrying fire debris directly into marine waters, or indirectly via river outflow. Post-fire landscapes where vegetation has been removed become highly susceptible to erosion, influencing both the secondary atmospheric transport of exposed soil particles into the ocean and the increased risk of landslides, runoff, and rapid mobilization of debris into coastal waters following intense precipitation events. This runoff pathway can take days to years to wash off fire-derived chemicals captured into soil and complex watershed networks.

Stimulating phytoplankton growth, following these fire inputs, potentially sets up a chain reaction on the marine food web. Starting with increases at the base and going all the way up the chain to impact larger marine organisms, such as the fish we eat to the mighty whales. This fertilization effect also has cascading impacts on the ocean capacity to store carbon, as phytoplankton die or are consumed, a small fraction of the carbon they used during photosynthesis ends up falling into the deep ocean. This loss of carbon to the ocean depths may even be balancing in part the quantity of carbon released during fires that was previously stored in soil and plants. While potentially significant, the marine fertilization effect of wildfires is a highly complex set of processes that depends not only on the type of fire and burnt source but also on the environmental conditions of the marine ecosystem receiving the inputs.

Wildfires are a natural and essential ecological process integral to the health, resilience, and sustainability of diverse ecosystems around the world. However, the increasing intensity and frequency of wildfires coupled with their growing intrusion into densely populated urban environments are introducing new environmental hazards. While natural wildfires play an essential ecological role by releasing nutrients that can stimulate marine productivity and carbon sequestration, urban wildfires incinerate man-made infrastructure, releasing hazardous pollutants into the environment. The narrative of wildfire impact on the ocean is therefore becoming more complex by adding an additional toxic footprint of heavy metals, microplastics, and persistent organic pollutants into the fire input mix.

The US West Coast, as one of the most productive marine ecosystems in the world, is a striking example of the complex impacts of escalating wildfires on ocean ecosystems. The coast, from Washington to California, is host to one of the largest Eastern Boundary Marine Systems (or EBUS) in the world: the California Current System. This vast oceanic region extends northwards beyond US borders to the North toward British Columbia, Canada, and to the South into Baja California, Mexico. Despite covering only 1 percent of the world's ocean area, EBUSs are biological super-reactor hotspots that sustain one-fifth of the global fisheries. The presence of coastal upwelling, driven by strong winds pushing surface water offshore and allowing deep, cold water to be brought to the surface, brings nutrients from below to the surface where it fuels abundant plankton life. Yet, the California Current System vibrant marine ecosystem now faces the counter-intuitive threat of intensifying wildfires along the US West Coast, potentially impacting the delicate balance that sustains teeming ocean life, biodiversity, coastal resilience, water quality, fisheries, tourism, and thus economy.

The catastrophic 2025 Los Angeles wildfires vividly illustrate the emerging complexity of wildfire impacts, highlighting both the potential ecosystem fertilization benefits from wildfire-derived nutrients and the severe environmental risks posed by urban-derived wildfire toxicants. As powerful hurricane-force Santa Ana winds pushed warm, dry air into the Los Angeles metropolitan region, the second largest city in the United States became primed for devastating urban wildfires to occur. Once ignited, vast clouds of smoke carried a cocktail of nutrients and hazardous pollutants offshore. Miles from the coast scientists aboard a research vessel caught under the plume described an acrid smell of smoke and burning eyes. They later recovered ash that had settled on the ocean surface as far as 100 miles offshore. Their plankton nets, typically teaming with life, returned not only zooplankton and anchovy eggs but also a troubling amount of charred material carried from the blaze. On land, coastal infrastructure burnt right up to the shoreline while debris piled up near rivers and creeks ready to be washed into already contaminated oily coastal surface waters.

As wildfires increasingly impact coastlines across the United States, it is imperative to understand their effects on already threatened marine resources. Addressing the emerging wildfire paradigm requires a fundamental shift in how we study, inform policy, actively manage, and respond to fire in a warming world. The complexity of modern wildfire regimes, including those in urban-influenced landscapes, necessitates interdisciplinary and transdisciplinary collaboration across fields such as ecology, climate science, public health, urban planning, atmospheric chemistry, and oceanography. Meeting this challenge calls for integrating long-term ecological monitoring with rapid-response sampling, advanced modeling, and satellite observations to better understand both immediate and long-term impacts. Recent work highlights the need to quantify cascading hazards such as the atmospheric and watershed delivery of pollutants, and the potential for toxic bioaccumulation in ecosystems. Wildfires can act as both nutrient sources and pollutant vectors, and only through holistic, collaborative research approaches can we anticipate these contrasting roles. This will inform regulatory framework, land-use planning, and debris removal management as well as public advisories. Such comprehensive efforts, when backed by coherent public policies, sustained investments, and adaptive management, are the clearest path to safeguarding public health, vital ecosystems, and economic stability in the face of this shifting fire regime.

Further Reading

Hamilton, Douglas S, Morgane M.G. Perron, Tami C. Bond, Andrew R. Bowie, Rebecca R. Buchholz, Cecile Guieu, Akinori Ito, Willy Maenhaut, Stelios Myriokefalitakis, Nazlı Olgun, Sagar D. Rathod, Kerstin Schepanski, Alessandro Tagliabue, Robert Wagner, and Natalie M. Mahowald. 2022. "Earth, Wind, Fire, and Pollution: Aerosol

Nutrient Sources and Impacts on Ocean Biogeochemistry." *Annual Review of Marine Science* 14: 303–30.

Tang, Weiyi, Joan Llort, Jakob Weis, Morgane M. G. Perron, Sara Basart, Zuchuan Li, Shubha Sathyendranath, Thomas Jackson, Estrella Sanz Rodriguez, Bernadette C. Proemse, Andrew R. Bowie, Christina Schallenberg, Peter G. Strutton, Richard Matear, and Nicolas Cassar. 2021. "Widespread Phytoplankton Blooms Triggered by 2019–2020 Australian Wildfires." *Nature* 597 (7876): 370–5.

Dr. Julie Dinasquet is with the Scripps Institution of Oceanography, University of California San Diego, La Jolla, California.

Dr. Douglas S. Hamilton is with the Department of Marine, Earth, and Atmospheric Sciences, North Carolina State University, Raleigh, North Carolina.

Expectations and Reality: Wildfire in the Wildland-Urban Interface

Dirch Foreman

Wildfire in the WUI is nothing new, but fires are different today. The reality of the WUI is that the fires are larger, the burning season is longer, the WUI is expanding, and urban type conflagrations are increasingly occurring. Fire will undoubtedly continue to impact the human-built environment. To better understand this, we must reconcile the public's expectations with reality.

The fire reality is evident. Fires are harder to control, they grow larger than ever before, and the fire "season" is now a year-round phenomenon. Global climate change adds to the intensity. Fire requires three properties to burn: "the Fire Triangle," fuel, oxygen, and a heat source. Fire does not discriminate between structures, landscaping, and natural vegetation as fuel. Oxygen is abundant in the world around us. Ignition sources are everywhere, from lightning, public utilities, and human actions. In nearly four decades of firefighting, I have witnessed the fire season, which historically only affected the "summer" months, grow by at least 80 days, to a continual marathon of firefighting. High winds, low relative humidities, high fuel loads, including structures, which are just trees in a different configuration, all lead to higher fire intensities, and rapid-fire growth. Fire suppression efforts are often quickly overwhelmed and become ineffective.

I believe there is an unrealistic public view of fire suppression. The expectation that large, complex wildfire can be overpowered with more people, more air tankers, more bulldozers, more technology is faulty. This assumption fails to recognize how complex large fire operations are and how close to extremely dangerous margins firefighters often operate, for the public. Successful control, under trying conditions, is marginal at best. The nation's firefighting organization, federal, state, and local, has more personnel and aircraft than the armed forces of many countries. There is no perfect number or formula for how many firefighters are needed and more aerial assets are not the solution. Firefighting alone cannot prevent WUI fire. The firefighters assigned to major complex fire incidents are not just from your local fire department. They come from all over the county and sometimes from other countries, to selflessly work to save property and protect lives. They perform this hard, dirty work for people they have never met, in a place where they don't live. The public cannot continue to blame firefighters for WUI conflagrations when a large portion of the fault lies with the public.

The WUI can be described as the place where wilderness meets human-developed land or developed land versus undeveloped, unoccupied land, as well as any number of similar definitions. The WUI has increasingly become the focal point of wildfire because of home destruction and threatened lives. Approximately one-third of homes in the United States are in WUI areas and the number of homes destroyed in WUI fires has nearly doubled over the last thirty years. It is hard for many people to accept, but this is not a wildfire problem, it is a people problem. About 80 percent of homes lost to WUI fire are due to rapid-fire spread, ember showers, and items outside the home, such as firewood piles, debris, patio furniture, wooden fences catching fire and spreading the fire to the home. It is time for the public to recognize that fire suppression is not the answer. Preventing home ignitions is.

There is nothing new to this view. In 2000, I was assigned to a fire in New Mexico where thirty-two homes burned. In each case, the house fire was caused by human- generated materials on the exterior igniting the house. The "tsunami" of fire rolling out of the forest is all but a myth. Structures ignite and then burn, largely due to lower-intensity ground fires and ember showers. The largest contributor to ground fire spread during the 1991 Tunnel Fire in the Oakland Hills, with over 3,000 dwellings destroyed, was homeowner landscaping. We see this time and again where development code and homeowner associations mandate flammable landscaping. The landscaping problem is further magnified by unenlightened homeowners. There is not a one-size-fits-all solution to WUI fire. However, homeowners can take definitive action to create a less flammable home. Not plastering the exterior walls with combustibles, trimming trees, raking yards, removing all combustible material within 5 or 10 feet of the exterior, placing metal screen on all vents, and keeping "stuff" away from the home will go a long way. Subdivisions can be arranged to frustrate fire spread. Supporting prescribed fire by not complaining about smoke, occasional escapes, and temporary ugly scenery will aid in keeping fires out of human development.

Unwanted wildfire will continue to occur. Fire will increasingly impact the human-built development. Not all fires will be overpowered by man and machine. Codes and HOA's allow the unenlightened to build and landscape in unsmart fashions. Some relatively simple actions can greatly improve the survivability of structures in the WUI. And remember, as one of my mentors would say, when you point a finger at somebody else, there are three fingers pointing back at you.

Dirch Foreman is the retired Chief of the Highlands Fire District in Northern Arizona. He currently serves with the National Wildfire Coordinating Group as Incident Commander 3, Operations Section Chief 3, and as Safety Officer 2.

Hot Mess

Jeff Shelton

To truly understand how we got here and, more importantly, how we move forward, you have to look back, and inward, and outward. Honestly, wildland fire is quite simple. It's fuel. It's weather. It's topography. It's also money. It's also politics. It's also people. People with widely diverse perceptions, experiences, philosophies, priorities, biases, and knowledge; all linked and filtered through a well-intentioned lens of "what right looks like."

Wildland vegetation is commonly referred to as "fuel." To be more precise, it's stored energy. The more energy available, the harder it becomes to manage. Nature, as a whole, and vegetation as a subset, will take full advantage of any opportunity to increase its influence. Fire is a physical force as natural as rain and wind. Vegetation has a love–hate relationship with fire. Understanding that the same destructive force is also a purifying mechanism, many species of vegetation have become "fire adapted." This biological détente allows the strong to survive and the weak to perish.

Unfortunately, in our attempt to be fair and good, our shared narrative on fire in the wildland now revolves around the word "bad." To that seemingly virtuous goal, we have congratulated ourselves over the course of the last century as we have gone to war with fire and mostly won. But, like a spoiled child, we have only made the situation worse. By refusing to recognize fire as a "green" process that incrementally released stored energy potential from the landscape, we have hit the limit of what can be managed. If someone were to walk up on a hillside near your community and place gallons of gasoline onto that slope, the response to law enforcement would be predictably swift. Yet that is exactly what we have allowed to happen year after year, decade after decade. Then we act surprised when fire ravenously feeds on this unnaturally dense energy source. The land is owed a fire debt and we are seeing the logical results of what happens when that debt goes to collection.

In a rush to rationalize our wildfire suppression and vegetation management strategies, our society has become irrationally focused on weather and climate. Though this makes us feel noble and moral, the harsh truth is that fire doesn't care about the why. Fire doesn't respect jurisdictional boundaries. Fire doesn't care about your legal rights. Fire doesn't care if you recycle. Fire doesn't care if you drive an electric vehicle. Fire doesn't care if you are a Republican, Democrat, or Libertarian. If you are reading this statement and you are feeling any form of anger at the audacity that the climate is less to blame than our policies, then you may be part of the problem. It is my contention that if we regularly managed vegetation to replicate the brutally elegant process of fire, we would not be facing a widespread community conflagration crisis.

The fire season is now the fire year. We are repeatedly seeing destructive fire activity in winter. Honestly acknowledging how the short- and long-term consequences of weather and climate supply a powerful accelerant to the collective mismanagement of our open space is the first uncomfortable step to implementing a practical solution to our dilemma. By admitting that well intentioned, but wildly useless policy is equally to blame as the climate, each side can sadly claim victory for their tribe and maybe we can move past the siloed mentality that only places blame on the other side. Unless we can accept that the perceptions and philosophies of our opposition have legitimate merit, we are doomed to continue to "love the land to death." Currently, each side only sees itself as the real stewards of the land.

When evaluating the "fire problem," we tend to focus on either fuels or weather. Meanwhile topography is regularly given a pass. Fire loves terrain. It enjoys the ability to align its upward convective energy with steep slopes. It adores the focused acceleration from canyons and valleys. It relishes the compressive heating and vertical chaos it creates when flowing downhill into a mountain wave. Riding a mountain wave is not as fun as it may sound, especially when it is entrained with fire. You know who else loves terrain? People. We do everything in our power to get a better view. It's in our nature. Getting a home with a view frequently comes with a view of nature. The more nature, the better. The better the view, the more expensive it is. The more expensive it is, the more exclusive it is. The more expensive it is, the more tax revenue it generates. The more expensive it is, the more it costs to rebuild when it predictably burns to the ground. In the classic "sunken cost trap" we keep approving permits and insurance keeps getting underwritten.

If you believe that the costs to rebuild fire-ravaged communities will only be absorbed by those communities, you are mistaken. Insurance is a collective. A fiduciary promise for all who contribute. Affordable insurance is contingent on the insurance companies having enough money to make a profit through investment. Payouts in the billions of dollars are crushing their long-held financial strategy. The new reality in wildfire insurance will be painful for everyone. Regardless of your location, expect to subsidize the expense of rebuilding with a view.

Leadership is the antidote to bureaucracy. Leadership is often substituted as a proxy for administration, management, or supervision. To the disappointment of administrators, managers, and supervisors, they are related, but absolutely not equivalent. A good administrator enforces rules, policies, guidelines, recommendations, procedures. A good leader performs almost no enforcement. They establish intent by providing clear context to task, purpose, and endstate. They communicate what they want done, why it needs to be done, and what it should look like when it's done. Rarely, do they tell you how to do it. This "mission driven culture" is inspiring for the individual, stimulating for the leader and healthy for the organization.

It took nearly a century of negligent interaction with the natural environment to get us painted into this explosive biomass corner where beneficial fire can often only responsibly be introduced in artificially short windows. However, our early hubris and relative effectiveness in suppressing fire have taught us valuable lessons on how to manipulate the fuels portion of the fire equation to replicate the successes of nature. Our ability to suppress wildland fire remains largely a function of the military structure and precision we have brought to bear against fire. Despite a volatile fire environment, the leadership, technology, and funding of incident command teams continue to produce impressive results. I contend that this same incident management team expertise will be the restorative solution to our lands. Imagine deploying complete incident management teams with the myriad of tools at their disposal, tasked only with treating prioritized, hazardous landscapes during times of maximum benefit.

Fire use teams are not a new idea. They were just never implemented at scale due to financial and staffing pressures. As such, they quickly fell victim to the increasing urgency of wildfire suppression. Now that wildfire is regularly wiping entire communities off the map, the decision to invest in multiple, full-time, year-round incident management organizations fully dedicated to the sensible reintroduction of fire as a normal part of the environment seems intuitive. With assistance and coordination from private commercial interests in the timber and grazing communities, real progress can be accomplished.

There are large areas of agreement between "the land is mine, I do what I want" contingent and the "humans are a contamination upon the land" faction. Strong leadership can bridge these apparently competing groups and provide a reasonable middle ground for change that both sides can advocate for. Fighting amongst ourselves is about as rational as fighting with nature. The choice is ours. Align our practices with the wisdom of nature's processes or continue to suffer the consequences. With time on its side, nature will always win. The sooner we realize this truth, the sooner we can clean up this mess.

Jeff Shelton is a Wildfire Analyst and Fire Captain in Southern California.

- - FIRE - -
DESTROYS GAME
IS FOUND IN
GREEN FORESTS
FIRE
DON'T
TURN HIM LOOSE
ON THE
NATIONAL FORESTS

4

Profiles

American history is punctuated with men (and a small handful of women) who have played key roles in the development of firefighting methods, fire science, the administration of agencies with responsibility for preventing and suppressing wildfires, legislators and officials who have shaped the laws, ordinances, policies, and regulations that have evolved to create the way we relate to wildfires. They represent fields such as forestry, ecology, biology, and history. They have served in leadership positions in the National Park Service and US Forest Service, while others are inventors, politicians, teachers, and trailblazers. Some of the individuals are well known for careers that span decades, and yet others are short entries in wiki biographies and scholarly journals about wildfires. Together, they tell us about the professions and people responsible for what we know about fire behavior, and how to protect our communities from wildfire's damaging power.

Stephen F. Arno (1943–2022)

It could easily be said that Steve Arno lived his life in the forest. At least that is how his mother's diary recorded his childhood, writing that he had loved hiking and spending time in the forest since he was two years old. At age five, he hauled big chunks of Douglas fir bark from the beach up to their house where his mother used the wood for fuel in their stove. He lived on Bainbridge Island, Washington, with his parents Siegfried and Grace Turner Arno until 1950, when

---

William Buckhout Greeley (1879–1955) served as the third chief of the US Forest Service from 1920 to 1928. Greeley made fire prevention and firefighting the agency's primary mission, in part through the passage of the Clark-McNary Act of 1924, which helped to bring states into partnership with the USFS to control forest fires (Library of Congress).

the family moved to Bremerton. At age eleven, he would capture logs using the family's boat, sawing and splitting them into firewood on the beach. From those early beginnings, Steve Arno studied the forest, worked cutting timber, worked researching and protecting timber from wildfires, raised two sons who worked in forestry, and he and his wife built a home on their own managed tree farm where he practiced restoration forestry.

Steve attended Olympic Junior College for two years after graduating from high school, and then transferred to attend Washington State University's Department of Forestry, graduating in 1965. During the summers between college, he worked as a ranger and naturalist at both Sequoia/Kings Canyon National Park in California and Washington's Olympic National Park. In September 1965, he married his wife Bonnie and they moved to Missoula, Montana, where he attended the University of Montana for the graduate degree in forest ecology, earning his PhD in 1970. He started work at a local lumber mill until it down sized, then began what would become a 25-year-long career with the US Forest Service at the Missoula Fire Sciences Laboratory. His primary research interest was the role of fire in shaping and sustaining forests in both Montana and the West. As a scientist, he experimented on how prescribed fire could be applied in combination with other fuel treatments to make native forests become more sustainable. He had a particular interest in the whitebark pine forests. As a fire historian, his interest in the evolution of forest fire policy in the United States grew out of his own work in fire ecology.

As a historian, Arno spent time in the Rocky Mountains studying fire-adapted ecosystems and the role of fire in the development of USFS policies. In that research, Arno credited a number of events as being responsible for the changing attitudes about fire within the forestry profession. He wrote about how the Big Burn fires of August 1910 fueled the debate over light burning for another two decades until the issue resurfaced at the end of the Second World War and the work of several academicians like Harold Weaver and Harold Biswell who supported the concept of what would then be called controlled burning. Arno was able to trace the environmental sensitivities of the 1960s and 1970s to the movement to return fire to the landscape for ecological benefits, the growth in development in the wildland-urban interface, and his belief that fire ecology needed to be integrated into firefighter training, a radical idea to some old-school members of the profession. He explained why by writing, "Firefighters who do not recognize fire as an integral component of the forest ecosystem may view their mission as a heroic attempt to save the forest, which may lead them to take inappropriate risks" (Arno, 2014).

Throughout his career, and long after his retirement in 1999, Arno was both a prolific researcher and writer, credited with authoring more than 250 papers, 100 peer-reviewed articles, and eleven books. He continued to work on the forest topics he cared about so deeply. In 2002, his book, *Flames in Our Forest: Disaster or Renewal?* explored how fire had been used throughout American history and how exclusion of fire from our forests had led to negative consequences. He co-authored another book, *Mimicking Nature's Fire: Restoring Fire-Prone Forests in the West* that is written in a more prescriptive tone. Advocating what he calls "a new direction in forest management" based on forests and fire ecology, Arno's use of case studies makes a compelling case for change in the direction and attitudes that colored much of the twentieth century's approach to fire. In 2020, another book chronicled what he called the untold story behind the region's Douglas fir from what might be called a literary perspective that extolled the specie's virtues.

Somewhat coincidentally, among his numerous awards was one named after the pioneer whose ideas he often wrote about, Harold Biswell. In 2004, Arno was given the Biswell Award by the Association for Fire Ecology. He was also given awards ranging from one bestowed by the Montana Wood Products Association to the Society of American Foresters. He and his wife continued to live on the family's uneven-aged ponderosa pine forest they had purchased in 1971 west of Florence, Montana, where they often gave tree farm tours to share their experiences. After Bonnie and Steve died, their two sons Matthew and Nathan continued to do so.

Robert Douglas Barbee (1935–2016)

Leadership in times of crisis has often been said to be the true measure of an individual. If that is true, then Bob Barbee's career with the National Park Service (NPS) is exemplary of that attribute. As superintendent of Yellowstone National Park from 1983 to 1994, he is best remembered for how he handled the epic 1988 wildfires in the park while confronted with opposition, derision, and calls for his resignation during the crisis. It is his leadership during the fires that the group The Coalition to Protect America's National Parks says ultimately shaped the future of federal wildland fire management policy.

Bob Barbee was born in Columbus, Ohio, and attended Colorado State University (CSU), graduating with an undergraduate degree in biology in 1958, and a master's degree in natural resources management in 1968. Later in life, he would be named as one of CSU's distinguished alumnus. He went to Fort Benning, Georgia, as part of the US Army ROTC program, and like many college students at the time, worked as a seasonal ranger in both Rocky Mountain and Yosemite national parks. His first permanent position with the National Park Service was at Carlsbad Caverns National Park where he worked as an interpretive ranger.

As is the case with many NPS careerists, he moved from park to park in quick succession in various capacities throughout the country: Grand Canyon, Point Reyes, Yosemite (again, as a fire ecologist), Cape Lookout National Seashore, Cape Hatteras National Seashore, and in higher level positions with the Western Regional Office in San Francisco as Chief of the Division of Interpretation. Next came positions as superintendent at Hawaii Volcanoes National Park, Redwoods National Park, and then at Yellowstone from 1983 to 1994.

It was as superintendent at Yellowstone, a position which he told the Associated Press was "the best job in the world" that he found himself at the center of one of the NPS's most historic controversies. At issue was the Department of the Interior's policy, within the NPS, of allowing naturally caused wildfires to burn, with the caveat that the fires not threaten life, property, or historic sites. Yellowstone is huge—2.2 million acres. There were eighteen lightning-caused fires that began on June 14, 1988, with the Storm Creek Fire, and eleven of them burned themselves out. Although park officials expected that July would repeat its historically wet weather pattern, it was drier than predicted, and no new naturally-caused fires were allowed to burn after July 21 except those adjacent to existing fires. After July 21, all of the fires were fully suppressed, including one, the North Fork Fire, that started the next day. By August 20, the total fire acreage had doubled to more than 480,000 acres; on September 7, a firestorm blasted the area around Old Faithful; and on September 11, rain and snow begin to fall, stopping the fires' advance. A total of forty-two fires were caused by lightning by year's end, with nine of them caused by humans. A total of 36 percent of the park was affected, with more than $120 million spent fighting the fires within the Greater Yellowstone Ecosystem, with a total of 10,000 people involved in various firefighting efforts (National Park Service, n.d.). No one died or was seriously injured in the park because of the fires.

Barbee had become defensive in his discussion of the fire's damage, noting, "Yellowstone Park is still all here. There will be a changed face to Yellowstone, but not a devastated face, not a destroyed face—that's just folly, that's myth. What there will be is just a different face." In an attempt to put a positive spin on what had happened, he added, "This event provides a great educational opportunity to see what a powerful force like fire can do, and how the land responds to it" (Associated Press, 1988).

The methods and policies that Barbee followed in managing the Yellowstone fires were controversial at the time, but considered mainstream now. Secretary of the Interior Donald Hodel called the firefighting policy a "disaster" and Montana Senator John Melcher, who was up for re-election, called for congressional oversight hearings about the way the fires started and how they were attacked. He was the subject of jokes that poked fun at his name, such as "What are the fires in Yellowstone called? Why they're called Barbee-cues." He referred to the conditions that he faced as ones that come together every 200 years or so, despairing "There was no way to anticipate this; it's never happened before" (Associated Press, 1988).

Barbee had long been considered a possible candidate for becoming the head of the NPS in the next administration, but the fires in Yellowstone changed all that. His last position with the NPS was as Regional Director of the Alaska National Parks, retiring in 2000. His 42-year career in public service included recognition with the Department of the Interior's three highest honor awards for Superior Service, Distinguished Service, and Meritorious Service, as well as awards from several conservation organizations. He and his wife of fifty-eight years, Carol, retired to Bozeman, Montana, where he passed away in 2016. His family included three daughters and seven grandchildren, and they shared his interest in mountain climbing, hiking, and the outdoors. Throughout his life, he was an avid photographer who studied with Ansel Adams in Yosemite, and he and Carol visited all seven continents where he photographed some of the world's most awe-inspiring natural areas. He died at his home with family by his side.

Harold Hubert Biswell (1905–92)

With the nicknames "Dr. Burnwell" and "Harry the Torch," Harold Biswell was both controversial and revered within the fire profession. He has also been called a pioneer in fire ecology, shifting the focus of fire culture with a passion that led his students and followers—some referred to themselves as "converts" to call him "Doc" with great fondness.

Born as the fifth of ten children to Charles Biswell and Ellie Rush Cornelius in Fayette, Missouri, Harold Biswell attended Central College, earning a degree in zoology in 1930. He went on to receive his PhD in plant ecology from the University of Nebraska in 1934. He then worked for the US Forest Service doing research on rangeland resources both in California and in North Carolina at the USFS Southeastern Experiment Station. In 1947, he joined the faculty in the University of California Berkeley's School of Forestry.

Combining his expertise in rangeland management with his academic credentials at Berkeley, Biswell began working with ranchers to improve their game range by using fire in California's chaparral vegetation. He tested some of his theories in Madera County's Teaford Forest in the foothills of the Sierra Nevada, showing farmers and ranchers how they could use fire to replace woody vegetation, which could then be replaced with grasses suitable for livestock. Although his range improvement burning proved popular, the practice declined after 1955 when more homes were being built on adjacent wildlands and ranchers feared they would be held liable for damage if a fire got out of control. Working with the US Forest Service and the Bureau of Land Management in Mendocino County, along with the California Department of Fish and Game, Biswell's planned burns were used to open up brushlands to increase the deer population as part of his wildlife habitat research. But as had been the case in Madera County, concerns about escaped fire caused the practice to decline.

In 1951, Biswell changed his research focus to the ponderosa pine forests in California's Sierra Nevada that were similar to the forests he had seen in Georgia. A two-part study was done adjacent to the Hoberg's Resort in the North Coast Range, and on the Teaford Forest near North Fork. He studied lightning strikes and what he term "frequent light burning" during aboriginal times, and compared forest cover after decades of fire suppression. In several subsequent articles, Biswell explained the techniques he used for broadcast burning and cleanup burning, clearly advocating the use of controlled fire. From then through the early 1960s, Biswell would be attacked by fire professionals and scholars alike for spreading what some termed "opinionated misinformation" about prescribed burning that they believed negated attempts at fire prevention. Today, his research is considered pioneering work is considered to have changed the trajectory of fire management among many federal agencies and among private landowners, bringing prescribed burning back to the management toolbox as a way of recreating the natural fire regime.

One of the agencies most affected by Biswell's experiments was the National Park Service (NPS), which benefitted from both his contemporaries' research and those of his former students. In 1962, the Secretary of the Interior asked Dr. Starker Leopold to head a committee to examine wildlife management in the parks, but the committee went further than that. They recommended that the national parks be managed as complete ecosystems, including the use of fire. Other scholars have noted that Leopold's report was no doubt influenced by his close association with Biswell, including times when Leopold attended Biswell's field day demonstrations of the use of prescribed burning. The Leopold report later was incorporated into NPS policy in 1968.

In 1965, Biswell's research agenda changed from the ponderosa pine to the giant sequoia in an area of California near Sequoia and Kings Canyon National Parks. The focus was on determining whether there was a difference in giant sequoia seedling survival between burned and unburned areas. His research showed that different species had varying fuel characteristics that would affect fuel behavior in studies that would continue to be replicated by other researchers. Not just a researcher, Biswell frequently conducted demonstrations of prescribed burning for educational purposes to help enlighten others about the use of fire.

Upon retirement from Berkeley in 1973, Biswell chose to continue to teach. He taught fire ecology at the University of California, Davis, along with a series of university extension classes. Other classes were offered at Yosemite National Park, at Calaveras State Park and Mt. Diablo State Park, and in San Diego. Courses like those Biswell taught are required for state park rangers, and influenced the restoration of prescribed fire in California's Department of Parks and Recreation, according to a former graduate student who worked with Biswell. Similarly, students who attended Biswell's field days and fire demonstrations from the California Department of Forestry and Fire Protection (CALFire) helped change not only their state's policy on suppression, but also the federal policy adopted in 1978 to encompass a fire management program that included prescribed burning. Biswell himself reported a ranger commenting, "In the fifties we were all making fun of Harold and fighting him. Now, 30 years later, we are all working for him" (Biswell, 1989).

Biswell's understanding of fire behavior and wildland fuels is best illustrated by his research into the weather conditions before the 1923 fire that destroyed

625 homes in Berkeley, California. He was convinced that if fuels were not treated, an event like the 1923 fire would reoccur, which happened in 1970 when thirty-seven homes were burned in the East Bay hills above Oakland and Berkeley, and again in 1991 in the Tunnel Fire that led to the death of twenty-five persons, injuring another 150. Although small in size at 1,520 acres, over 3,200 structures were destroyed at an estimated $3 billion in 2024 dollars.

Over the course of his career, the prolific researcher and teacher authored 170 professional papers on topics related to fire and rangeland improvement, and his book *Prescribed Burning in California Wildland Vegetation Management*, published in 1989, became a classic. Biswell received a Fulbright Award and a Guggenheim Fellowship, along with the designation of Professor Emeritus from the University of California, Berkeley. Societies named awards after him, and the California Licensed Foresters Association created a scholarship in his name for graduating high school seniors in California interested in a forestry career. "The Biswell Symposium: Fire Issues and Solutions in Urban Interface and Wildland Ecosystems" was held in his memory two years after his death in 1992 to honor the pioneering individual who had influenced, in some way, nearly all of the 350 participants who attended.

Coert DuBois (1881–1960)

Known as "the grandfather of forest science," Coert DuBois played a key role in the development of California's firefighting units and introduced management techniques that led to a remarkable change in attitudes toward wildfires. DuBois was the fifth child of Evalina Patterson Kimball and Dr. John Coertland DuBois, from whom he inherited his name. After graduation from high school in 1899, DuBois attended the exclusive Biltmore Forest School on George W. Vanderbilt's Biltmore Estate near Asheville, North Carolina. The estate was also home to the 125,000-acre forest laid out by landscape architect Frederick Law Olmsted and managed by French-trained American forester Gifford Pinchot. Founded in 1898 by one of Germany's leading foresters, Carl Schenck, as the first school of forestry in North America, DuBois was one of the first Biltmore Forest School students to enroll in the year-long "mounted school" program. The curriculum included horseback riding and courses ranging from road building and lumber processing to silviculture and sustainability. Unlike the majority of American universities, Schenck's program was based on the German model of hands-on application of theoretical models rather than textbooks. After financial disputes with Vanderbilt, Schenck was fired from his position in 1909, and in 1913 the school closed its doors, having graduated 360 men including Coert DuBois.

Along with several other Biltmore graduates, DuBois joined the US Forest Service in April 1900, where Pinchot had been named the agency's first chief, as a student assistant in Washington, DC with a salary of $25 a month. In 1904, he moved on to an administrative position determining the location and extent of the forest boundaries, and a year later, was named inspector for the Rocky Mountain and Southwestern sections. In 1908, he was appointed Associate District Forester for California, and three years later, became the United States Forester in charge of the district.

DuBois focused on the protection of California's young forests and the timber supply the state needed for agricultural and industrial development. He expressed his concern over the forest fires that had been wantonly set or started by what he termed "sheer carelessness and indifference." He opposed light burning, where the undergrowth and downed logs in a forest were routinely set on fire in the spring and fall with the belief that there would be no damage to mature timber, thus making forests immune to wildfires during the summer. He rejected the theory that California's forests had persisted for centuries despite lightning fires and those set by Indians and early settlers. Under DuBois, California foresters took on both timber companies and fire itself, introducing new methods such as firebreaks, requiring loggers to pile brush and debris to reduce fuel loads, and requiring the companies to prevent and extinguish forest fires. Adhering to the gospel of efficiency in vogue at the time, the California districts also established a system of lookout towers and ranger patrols. After the 1910 fire season ended, DuBois admonished his staff, noting in one Forest Service document, "Until we can handle fire on the Forests entrusted to our care, we cannot practice forestry on them … to protect them, is our first duty to our employers—the people of the United States … It's time we got war-like—time we put a fighting force on our frontier to stand over the enemy with a club" (USDA Forest Service, 1910). His pioneering forest protection plan included paying trained standby fire crews, giving rangers fire assistants, urging the state to require burning permits during fire season, requiring campfire permits for forest visitors, and developing public relations fire prevention messages, using, among other media, motion pictures. DuBois approached fighting fires in the technical, systematic way that had become part of the nation's Progressive movement. He required better communication equipment and mapping of forests, saying that wildfires were not an act of God, but a human-caused problem that could be managed and controlled. The results of his work were published in 1914 in the volume, *Systematic Fire Protection in the California Forests*, in which he called for coordination and cooperation of fire management from a more scientific perspective. A year later, he circulated a memo that morphed into the "let burn policy" that allowed non-timbered government land to burn if the cost of fighting a fire was disproportionate to the value of the timber involved.

Responding to President Woodrow Wilson's call to action, DuBois voluntarily entered military service in 1917 as a major in the US Army's 10th Engineers Forest Regiment, American Expeditionary Forces, in France. Upon his return from the war with the rank of lieutenant colonel, he was appointed to the Consular Service in 1919, serving in Washington, DC, Paris, India, and Italy. In 1942, President Franklin Roosevelt appointed DuBois to the Anglo-American Caribbean Commission. DuBois died in 1960, and his wife Margaret in 1969. Their two children died in a tragic suicide pact in 1935.

Harry Thomas Gisborne (1893–1949)

Wildfires have been part of the American landscape for thousands of years, but it was not until the 1920s that a young man from Vermont became one of the first to approach forest fires from a scientific perspective. Although there are many photos of him in his official US Forest Service uniform, he is just as likely to be pictured with one of his many inventions or scientific instruments that he used to research fire risk factors, fire behavior, or fire retardant.

Harry Gisborne was born in 1893 to a Montpelier, Vermont, family active in the lumber business, and at any early stage, he became interested in forestry. A 1917 graduate of the forestry program at the University of Michigan, he once said that he chose forestry as a profession because he loved the woods and hills and wanted to work outdoors. For a short time, he worked as a timber cruiser in Oregon before serving with the Tenth Engineers during the First World War. He then joined the US Forest Service (USFS) and was assigned to Missoula, Montana, as head of the Rocky Mountain Research Station. While there, he conducted numerous studies at the Priest River Experimental Forest in northern Idaho. It was at the research station that he established himself as a forest fire scientist, a career label that he had never heard of while attending a summer forestry program in Vermont in 1911. In 1922, Gisborne was appointed as a forest examiner and director of the Priest River station at a salary of $1,920. The land on which the station was located was particularly attractive to the USFS because all of the major trees found in the Northwest grew there in the same proportion as on similar land in the region. In addition, there was a burned over area where the staff could develop a tree nursery as part of the agency's emphasis on silviculture and its Department of Silvics.

Gisborne was especially interested in climbing what was called the Weather Tree—a 150-foot-tall larch that had platforms built into it that allowed him to study the wind's behavior from different levels of the tree. He was interested in what would emerge as the science of fire behavior, and eventually, the development of the National Fire Danger Rating System that was first implemented in 1934. He had what writer Norman Maclean called a childlike interest in lightning that led to observations and then predictions about fires. Known by his nickname Gis, his students as well as colleagues variously referred to him as outspoken, sarcastic, hard-driving, and demanding. He self-described himself as the burr under their tails that motivated them to do their best for him and for science.

During his tenure with the agency, he gained a reputation as someone who had an ability to take devices that could be engineered and used to determine how forest fires could be predicted and prevented. He was proud of his ability to use new equipment such as the double tripod heliograph, the Asman aspiration psychrometer, and the blinkometer. One of his interests led to a built fire tunnel that could be used to burn fires under controlled conditions, taking into account factors such as wind speed and fuel type. It is believed that he had such a tunnel built in the basement under his office in the USFS Regional Headquarters in Missoula, Montana. He is also remembered for his desire to make sure that his research was not just scientific, but also that it had a practical application, with Maclean saying that his projects aimed to save from fire as many board feet of lumber as possible.

Among the honors Gisborne received was the US Department of Agriculture's Superior Service Award, becoming the first person in the Northern Rockies region of the USFS to be given the award. He died on November 9, 1949, while investigating the August fire at Mann Gulch, Montana, that killed thirteen smokejumpers. While walking through the area where the fire had burned in an attempt to learn more about what had happened, he collapsed and died of a heart attack. He had left instructions in his desk drawer saying he wanted his ashes spread on a mountain near the Priest River. Gisborne Mountain was named in his honor, and in 1951 a brass marker named him "Pioneer in Forest Fire Research."

William Buckhout Greeley (1879–1955)

"Why should we let our forests burn up while arguing over who should control the manner of their use?" (World Forestry Center, 2016). This is a question Bill Greeley asked during the tumultuous yet formative early years of the American Forestry Association and the US Forest Service. As a well-educated forester who also served in the US Army during the First World War, he elevated firefighting as the primary mission of the agency.

Born in Oswego, New York, to Frank Norton Greeley and his wife Anna Cheney Buckhout, Bill Greeley graduated from the University of California, Berkeley in 1901, and then attended the recently opened Yale Forest School in New Haven, Connecticut, receiving a Master of Forestry degree in 1904. The dean of the school, Henry S. Graves, wrote of Greeley, who was first in his marks, "He is a special star and I recommend him for almost any work that comes along" (Forest History Society, n.d.c). The Yale program was established by Graves and Gifford Pinchot, who would become the first chief of the US Forest Service (USFS) in 1905. After graduating from Yale, Greeley became a proud supporter of the Forest School, organizing the school's graduate advisory board in 1905, and becoming the first president of its alumni association. He was honored with the Yale Medal in 1955, and the William B. Greeley Memorial Laboratory was named for him after his death later that year.

In summer 1907, Bill Greeley married Gertrude Jewett, who he had met at Berkeley as an undergraduate. While Bill attended graduate school at Yale, she worked as a public school teacher in California. For a while they lived in New Haven, then moved back to California and eventually settled in Washington, DC. They raised three sons and a daughter. Son Arthur Greeley would also graduate from the Yale Forestry School, joining the USFS in 1935. He became a regional forester for Alaska, the regional forester in Wisconsin, then deputy chief and associate chief in the USFS Washington Office.

Bill Greeley quickly worked his way up through the ranks of what was initially the Bureau of Forestry (later named the USFS), assisting Pinchot and eventually was promoted to the agency's Washington Office. He was a district

Forester in 1910 during the Big Burn fires that destroyed thousands of acres of forests in Montana, where eighty-five men perished. At that transformative event, he would write, "The conviction was burned into me that fire prevention is the No. 1 job of American foresters " (World Forestry Center). Greeley took a utilitarian view of forests, believing they could be managed and that there was room for improvement in the timber industry's practices, a view counter to that of Pinchot. He began to advocate for compulsory slash burning, telling Americans they needed to change their wood burning habits, and sought more firefighters and cooperation between landowners and fire agencies.

Much like other men of his age and experience, he was commissioned as a US Army Major when the First World War began, assigned to the 10th Engineers (Forestry). His service took him to France in August 1917 where he became chief of the Forestry Section of the 20th Engineers and Attached Service Troops. In that position, he supervised 21,000 men operating ninety-five sawmills throughout France that produced 2 million board feet of lumber per day. He was awarded the Army's Distinguished Service Medal, the Distinguished Service Order from Great Britain, and named a Chevalier of France's Legion of Honor. He returned to the United States in July 1919 with his commission as lieutenant colonel in the Reserve Corps, and was often referred to as Colonel Greeley.

In 1920, he became the third chief of the Forest Service. Among his accomplishments during his eight-year tenure in the position was passage of the Clark-McNary Act of 1924, legislation that expanded the 1911 Weeks Act but which had failed to provide monies for actually fighting fires. Under Weeks, states that established forest protection agencies could apply for matching funds including money to pay for fire patrolmen's salaries, but not equipment. In 1920, Greeley asked Congress for a $1 million appropriation to encourage cooperation with the states in forest fire protection and forest renewal. He continued to lobby Congress to recognize the need for cooperation, including his recommendation that a national association be created to coordinate the activities of state foresters.

The first two sections of the Clark-McNary Act authorized the USFS to recommend and develop "an adequate system of forest protection and fire prevention in the several states" including financial assistance. Additional funds were added to the prior provisions under the Weeks law, providing $2.5 million to implement the somewhat vague language in the legislation that gave the USFS additional power with little additional oversight. Greeley was successful in his efforts to bring the states into a cooperative partnership to control forest fires, a policy that had not been undertaken before his leadership.

With the Clark-McNary Act fulfilling many of his professional aspirations, Greeley resigned as Chief Forester in 1928 to accept a position with the West Coast Lumbermen's Association, a timber industry trade group that had been incorporated in 1911, becoming the West Coast Lumber and Building Materials Association in 1964. He later was named Chairman of the Board of the American Forest Products Industries, Inc., another trade group that formed in 1932 with a focus on fire prevention outreach and self-regulation of privately owned forests. The organization evolved into the American Forest Institute after Greeley's death while retaining its mission of improving the management of American forests. He died in Suguamish, Washington, in 1955, and Gertrude, who had often worked side-by-side with her husband, passed in 1970.

Weldon Heyburn (1852–1912)

One of the most vocal critics of the US Forest Service, and more specifically, the agency's wildfire policies, Weldon Heyburn made a name for himself as a prominent attorney in Pennsylvania before moving west to pursue his law career and taking part in Idaho's mining boom. His parents, both English Quakers, were Sarah Gilpin and John Brinton Heyburn. They settled near Chadds Ford in Delaware County, Pennsylvania, where he was born and attended public schools and the Maplewood Institute in Concordville. One of seven children, he was schooled in part by private tutors. He attended the University of Pennsylvania to study civil and mining engineering, metallurgy, and geology. He also had an interest in law, studying under Edward A. Price, and was admitted to the Pennsylvania bar in 1876. He opened his law practice in Media, Pennsylvania, where he worked for two years.

Attracted by the mining boom flourishing in the West and influenced by Albert Deane Richardson's 1869 book *Beyond the Mississippi*, he moved to Leadville, Colorado, in 1878, and in winter 1883, moved to Shoshone County in the Idaho Territory, opening a law practice in Wallace. He staked several mining claims based on the knowledge he acquired in college, and in 1886, was elected to the Board of Directors of the Sierra-Nevada Consolidated Mining Company. Politically active, Heyburn was a delegate at the meetings that framed the state of Idaho constitution in 1889, a year before the territory became the 43rd state admitted to the Union. He also became active in the territory's Republican party, attending his first national party convention in 1888. He ran unsuccessfully for

Congress in 1898, and after statehood was achieved, asked to be considered as Idaho's first US Senator. In 1903, when Heyburn won his Senate election, he married Gheretein Yeatman (1854–1934), a young woman he had courted while in Pennsylvania. His wife had been previously married, and had lost a daughter at just one year of age.

After a stalemate in the legislative caucus in 1903, he was successful in winning enough support to gain the votes of the Republican-controlled state legislature, which elected him to his first term. At the time, US senators were not elected by the voters, and were heavily politicized by Idaho's mining industry. He won re-election in 1908, becoming an outspoken member of the Senate and a fierce protector of Idaho's natural resource industries.

Heyburn's permanent home was in Wallace, Idaho, one of the mining communities partly destroyed by a massive wildfire in 1910 and chronicled by Timothy Egan in his book, *The Big Burn* (2009). Somewhat ironically, Heyburn had spent much of his career assailing the USFS and the creation of the forest reserve system that surrounded his home. He became one of the opposition leaders against the agency. He called the forest reserves "an expensive, useless burden to the public," Egan writes, and later accused the 1910 fires that killed seventy-eight men of being due to public ownership of the forests. He took umbrage at anything natural resource-oriented supported by the president and by Gifford Pinchot. The Forest Service, he said in a speech against the agricultural appropriation bill that would fund the agency, was unjust, and an encroachment on states' rights, calling foresters "grafters."

Heyburn was the stoutest man in the Senate, requiring the Senate carpenters to build a special chair for him to sit in. While delivering a speech on the Senate floor in March 1912, he collapsed in front of his colleagues. His health declined over the next few months prior to his death in Washington, DC, on October 17, 1912 at the age of sixty. He and his wife are buried in the Birmingham-Lafayette Cemetery in Chester County, Pennsylvania.

The *Spokesman-Review* referred to Hayburn as leaving a "thorny legacy on natural resources," noting "History doesn't shed a kindly light on US Senator Weldon Heyburn," calling him "an irascible man who once stopped an orchestra in mid-performance because he didn't like a song it was playing" (Kramer, 2010). The song, "Dixie," was being played to a visiting member of Congress, and Heyburn explained the event was a Republican meeting and he wanted no such tunes. He was also said to have denied a young constituent a debate prize, noting "he does not seem to have learned enough to be a Republican." But Heyburn's legacy can be measured in other ways, such as the state park named in his honor (despite the fact that the senator wanted a national park, not a state one). Or the naming of Mount Heyburn in the Sawtooth Mountains, or the city of Heyburn, Idaho.

On the legislative front, Heyburn can be credited with introducing the bill that would become the Pure Food and Drug Act of 1906. The law was signed by Heyburn's nemesis, President Theodore Roosevelt, but more importantly, it was enacted when nearly 100 similar bills had failed to pass. This important law required that food, patent medicine containing alcohol, and drugs, which had been unregulated, be appropriately labeled, and prohibited manufacturers from adulterating or misbranding these products traveling through interstate commerce. Although the law did not create the current Food and Drug Administration, it is considered a foundational document, with partial credit due Weldon Heyburn.

Susan J. Husari (1953–)

The use of the term "trailblazer" would seem an obvious choice, since it refers to someone who is the first or the leader of a group. Add to that term words such as "hot shot" and "fire manager" to describe someone's career, and you would get closer to the profile of Susan Husari, who retired from the National Park Service in 2012. She held several titles in her 45-year career, and serves as a role model in a field where women still make up less than 10 percent of the workforce.

Sue is a native San Franciscan who often went backpacking in the Sierra Nevada with her dad. She says she was never a natural athlete, but she could hike, carry heavy stuff, and was familiar with the outdoors. She started her career as a seasonal firefighter, but planned on attending graduate school after graduating with a bachelor's degree in botany from Humboldt State University in northern California. She expected to have a career in the field of mycology (mushrooms) but one season as a firefighter led to another, and then another, and another. Firefighting, she says, is a pretty addictive occupation, especially when actually fighting a fire. She was part of the Happy Camp Brush Disposal Crew, and in 1975 at age twenty she joined the Lassen Hotshots, the first woman to join the crew and one of the few women hired to serve as a hotshot in the state. She says she still is not sure why they hired a woman—unusual not just now but even more so then. At a time when women were still trying to breach the barriers of occupations dominated by men, she says she was not treated differently than the men she worked with, except that she felt her coworkers often looked out for her. The majority of them were older than she was, with service during the Viet Nam war. While she did not have the same experience they did, she felt she was well-suited for the job as a firefighter because her super power was her endurance. She worked for two seasons on the hotshot crew, and says it was the best way for her to not only learn about fire, but also that it formed the basis for her understanding of fire ecology. She also worked as a helitack crew member and engine operator on the Lassen National Forest from 1976 to 1981.

She left firefighting to work for the Bureau of Land Management as a botanist for a few years before going to work at Pinnacles National Monument (established as a national park in 2013) and at Joshua Tree National Monument. From California she went to Everglades National Park in Florida from 1987 to 1991 first serving as fuels manager and then as the fire management officer. While in Florida, she helped develop the park's prescribed fire program in a region where the practice was much more accepted than in the rest of the country. She also was responsible for fire monitoring program while at Everglades. From 1991 to 2002, she held positions with the US Forest Service, including serving as Deputy Director of Fire and Aviation Management, heading up the Fuels Management Program as Assistant Director. These positions, in the USFS Pacific Southwest Region, brought her into contact with leading researchers from around the country, as well as fire managers from numerous federal, state, and local agencies. Her name appears in dozens of scientific and agency publications where her expertise in large-scale prescribed fire was informative.

Husari returned to the National Park Service in 2002, becoming the Regional Fire Management Officer for the Pacific West Region, retiring in 2012. This position involved the oversight of forty-four parks with fire management plans. Her career with fire did not end with her retirement, however. She served as a public member of the California Board of Forestry and Fire Protection, chaired the Resource Protection Committee, and co-chaired the Effectiveness Monitoring Committee for the board.

In discussing the role of women as role models, Husari says that she did not have any other women to model in the firefighting profession when she started so it took several years for her to learn how to be a leader in a male-dominated organization. She developed self-confidence, used her sense of humor, and was able to serve as a mentor other employees and the future leaders within the agencies where she worked. She saw many changes within the fire profession during her career, and noted that the agencies have become more inclusive in decision-making and less militaristic. Human factors, such as incorporating diverse points of view in making decisions and analyzing risk, are now integrated in firefighting and leadership. But those same factors are also what she considers to be the worst changes she has seen in her career. Decisions in firefighting, she says, have become highly political. This forces leaders and agency administrators to consider the political risks they are taking in the implementation of fire management. That can make people risk-averse in doing the right thing for the parks or forests that they manage. "The willingness to spend huge amounts of the taxpayers' money on some suppression efforts with political considerations in mind has been something that has been disturbing," she says (Lawatch, n.d.).

Abigail R. Kimbell (1953–)

One of eight children, Abigail Kimbell was the daughter of Arthur R. Kimbell and Mona. L. Jacob and part of a family that loved the outdoors. Growing up in Vermont, she spent time hiking, fishing, and camping in the White Mountains of New Hampshire, joking that as a child she endured what she jokingly calls "forced marches to bug infested lakes" to go fishing. She attended high school at Bellows Free Academy in St. Albans, Vermont. As her family moved throughout the region, she expanded her passion for natural areas into a career as the first woman chief of the US Forest Service. At the time, she supervised an organization of over 30,000 employees and an agency budget of nearly $5 billion.

Kimbell attended the University of Vermont, graduating in 1974 with a degree in forest management, and worked as a seasonal employee before joining the Bureau of Land Management in Medford, Oregon, in 1974. She received a master's degree in forest engineering from Oregon State University in 1982, where she met her husband, Edwin Stryker. She joined the US Forest Service as a pre-sale forester in Kodiak, Alaska, in 1977. Like many of those in the forest profession, she had numerous professional assignments, moving to Kettle Falls, Washington, to the Colville National Forest from 1985 to 1988, and to the Wallowa-Whitman National Forest in eastern Oregon from 1988 to 1991. She returned to Alaska in 1992–7 as forest supervisor of the Tongass National Forest until 1997, and then Wyoming's Bighorn National Forest from 1997 to 1999. From there, she held positions as the forest supervisor for the Pike and San

Isabel National Forests and the Comanche National Grassland in Colorado, and the Cimarron National Grassland in Kansas.

Her leadership potential was recognized in May 2002 when she was named the Associate Deputy Chief for the National Forest System lands in the agency's main office in Washington. This assignment placed her in an important position in the implementation of the Healthy Forests Initiative and the drafting of the Healthy Forests Restoration Act of 2003. The development of these policies not only included extensive negotiations on hazardous fuels reduction and forest thinning projects, but also numerous proposals related to wildfires. As initially released in 2003, legislation called for hazardous fuels reduction on 20 million acres of USFS and BLM lands near communities in the wildland-urban interface (WUI). In December 2003, she was named Regional Forester for the Northern Region in Missoula, Montana, responsible for the development and implementation of community wildfire protection plans in Idaho, Montana, and North Dakota. She was named Chief of the USFS from February 5, 2007 to July 2, 2009, retiring July 31, 2009. On the day of her appointment, Senate Finance Committee Chairman Max Baucus (D-Montana) sharply criticized Kimbell, saying she was "inclined to raise fees, close campgrounds and otherwise make it harder for people to access their lands to raise revenue," according to the *Washington Post* ("Senator Objects" 2007). Although she was the first female chief, she was followed in 2018 by a second woman, Vicki Christiansen, who left in 2021.

The *Congressional Record* (2009) credits Kimbell with a number of achievements, including improving firefighter safety and fire suppression cost containment, one of many controversial elements of her administration. In August 2008, she ordered forest service officials to cut their operations budgets to save $400 to $700 million to cover the cost of fighting fires. In the 1.5 million-acre Arapaho and Roosevelt national forests and the Pawnee National Grassland, officials trimmed nearly $900,000 from their $20 million budget. The budget cuts represented transfers of funds just before the end of the federal fiscal year, a time when the agency has already expended all or most of its funding on routine operations. As Laura Paskus (2008) noted in *High Country News*, the USFS had already received $332 million in emergency supplemental funding, but the cost-saving measures included mostly non-related activities such as research that was halted, forest planning projects that were discontinued, and employee travel and training that were denied.

Her critics believed she was pro-timber, but also note she was the person responsible for the largest reprisal action ever taken by agency whistleblowers. In 1992, her predecessor was alleged to have conducted illegal timber sales and sweetheart concessions to favored timber companies, among other wrongdoings. In 1994, eight Bighorn National Forest employees wrote a letter to the regional forester alleging the creation of a hostile work environment and mismanagement of the forest. Kimbell authored an agency reorganization that purged most of the whistleblowers who had not already faced reprisals or intimidation, and thirty of forty-four employees who had raised concerns were purged from their jobs. Kimbell claimed the reorganization was due to budgetary restraints, but between 1997 and 2003, environmental experts were replaced with office workers. Eight of the fired employees were represented by the Government Accountability Project in a lawsuit under the Whistleblower Protection Act. And in 2003, the US Office of Special Counsel supported the allegations. The employees were awarded damages of $200,000 divided among them, and the charges against Kimbell were dropped.

In 2009, Kimbell was named one of the outstanding alumni of the University of Oregon as one of the College of Forestry graduates from across the natural resources spectrum who have made outstanding contributions to the professions and/or programs of the college. Kimbell's life came full circle in 2021 when she was named to the Board of Advisors for the Rubenstein School of Environment at the University of Vermont, where her academic interest in forestry had begun in the early 1970s. She and her husband live in Central Oregon near where she once worked and where they continue to enjoy recreating on the national forests.

Edwin Vaclaw Komarek, Jr. (1909–95)

The son of a Czech immigrant, Frederick (Bedrich) Komarek who came to the United States in 1904, and his wife Stella (Stasie Hlavka), who was born in Illinois of Czech parents, Edwin's birth name was Zdenek when he was born in Chicago in 1909. The changing of names to Americanize or Anglicize an individual from Europe was common among immigrants seeking to assimilate to become part of their new country. He grew up in suburban Oak Park, Illinois, and the landscape was open prairie and forest. He joined the Boy Scouts and served as a volunteer when there were fires in the nearby preserves, influenced by the fire exclusion movement of the time. After graduating from Oak Park High School, he received scholarship assistance from the Chicago Academy of Science that allowed him to attend the University of Chicago to study zoology. Initially, he became a museum mammalogist at the Chicago Academy of Sciences, a field of study that led him to the South. He conducted the Smoky Mountains Faunal Survey alongside his younger brother Roy V. Komarek, and began authoring articles in scientific journals. The region was also where he met the noted wildlife biologist Herbert L. Stoddard, Sr. in 1934.

Stoddard had been hired by the US Bureau of Biological Survey to study the bobwhite quail in the longleaf pine-grassland forests of the Red Hills area between Thomasville, Georgia and Tallahassee, Florida. The area attracted wealthy visitors who bought up large swaths of acreage to use for hunting and building large residences called plantations. Although the out-of-state owners were often absent from their palatial homes in the summer, they supported efforts to determine why the resident quail population was declining. In 1924, Stoddard began the Cooperative Quail Study Investigation, headquartered at his own home, the Sherwood Plantation, with Komarek later serving as his assistant. Stoddard fought for the reintroduction of fire in the longleaf-grassland system, and he was among the first to advocate for ecological diversity in cultural landscapes.

Komarek began conducting research and living in Thomasville, Georgia, home to the Birdsong Plantation, which had been known as the Dickey-Birdsong Plantation. The 565-acre property, originally a longleaf pine forest, had been cleared for agriculture, acquired by Shadrach and William James Dickey between 1836 and 1872. The main house was built by William Dickey in 1912 as a working cotton plantation with a patchwork of open fields and woodlands. It was also here that Komarek married his wife, Elizabeth (Betty) Hester Barker, in 1935, and where they raised two children, E.V. Komarek and Betsy Komarek. Ed and Betty bought the Dickey Plantation in 1938 and began redeveloping it as a center for wildlife and soil conservation. The plantation was best known for Komarek's use of prescribed fire (known as control burning at that time) to reduce brush and restore native ground cover, sharing his knowledge of the role of fire with educational and youth groups as well as wildlife scientists. Birdsong became known as "the world's best birding window" with 164 species seen there.

Under his leadership, Komarek embraced using fire as the conscious development of habitat variety for wildlife diversity. Originally used for sharecropping, the open fields that were once used for growing cotton and later, for grazing cattle, were burned systematically as a way of attracting wildlife. With the assistance of the Quail Investigation researchers, Komarek conducted field trials using fire to study wildlife–plant interactions which would later become the groundwork for the new field of fire ecology. Roy Komarek became the head of the Birdsong Plantation in 1944, publishing numerous articles on fire ecology—a passion he shared with Ed.

From 1945 to 1988, Ed served as the general manager of the Greenwood Plantation, just west of Thomasville, studying the management of quail, timber, and management of the property. The main house was built in 1838, situated on 5,200 acres of forest land used for quail hunting, and 1,000 acres of old-growth longleaf pines, with some estimated to be up to 500 years old. The site became a major learning center for university groups, with Komarek offering field lectures to thousands of students and researchers involved in natural resource management. Greenwood also served as a retreat for politicians and celebrities, from President Dwight D. Eisenhower to Jacqueline Kennedy.

In 1958, Komarek founded the Tall Timbers Research Station, funded by Henry L. Beadel, who had also been involved in the Quail Association. Located in Leon County, Florida near Tallahassee, the two-story house was built around 1895 by Edward Beadel, a wealthy architect from New York who came south to hunt during the winter. Edward's nephew, Henry, was a wealthy philanthropist and naturalist who like Komarek believed in using fire as a land management tool. Henry Beadel bought the property in 1919, living there with his uncle from 1921 to 1963. He left his entire estate to be used as an ecological research station.

Beginning in 1962, Komarek used the site for the Tall Timbers Fire Ecology Conference series. The programs became the first professional meetings in the United States to deal with the beneficial effects of beneficial burning on natural resources. Today, the Tall Timbers Fire Ecology Program, according to the group's website, is recognized as the home of the study of fire ecology, and as an advocate to protect the right to use prescribed fire for land management. The station's research focus remains the ecology and management of fire-dependent ecosystems.

As Komarek's reputation in the field grew, he traveled widely in search of Tall Timbers speakers while lecturing throughout the United States and more than a dozen other countries. Birdsong's Nature Center began operating in 1981, and was established as a non-profit organization in 1986. The property was later named to the National Register of Historic Places.

Komarek was a scholar, publishing more than eighty articles in academic journals and receiving an honorary Doctor of Science degree from Florida State University in 1971. But he was also known as a practitioner, serving on the Georgia Game and Fish Commission and organizing fire ecology conferences. His brother Roy, also a well-known author on fire ecology, died in 1989. Ed died in 1995, and his ashes were buried at Birdsong; Betty, also a scientist and educator, passed in 2002.

The Nature Conservancy

Often known by its acronym, TNC, The Nature Conservancy has a broad mission that is usually associated with the equally broad environmental movement, although when the organization was founded in 1951, it was well ahead of the curve of that growing consciousness. Now active in seventy-nine countries around the globe, with more than a million members, TNC is considered one of the world's leading non-profit organizations working on issues related to wildfire. Since 1962, when the group conducted its first controlled burn at the Helen Allison Savanna in Minnesota, TNC has expanded its interest beyond the protection of diversity. The agenda is now one that integrates wildfire into subjects as broad as diversifying the fire management workforce to supporting indigenous fire practices.

With a goal of bringing the role of fire in nature back into balance and helping communities prepare for, manage, and live safely with fire, TNC has programs operating on numerous landscapes and with diverse populations. One of the largest programs deals with prescribed fire. The average burn per year conducted by TNC is more than 112,000 acres. In addition, they help partner groups and agencies burn about 287,000 acres each year in the United States, with many years exceeding 350,000 acres. There are about 500 qualified prescribed fire staff and volunteers who participate, trained using the same wildland fire requirements used by federal agencies in the United States. Its own standards for qualifying prescribed fire personnel actually exceed those promulgated by the National Wildfire Coordinating Group, made up of all federal and many state fire management agencies. TNC says no other private organization can match its fire management capacity and experience; it is the only non-governmental organization in the federal fire qualifications system.

TNC has strict requirements for becoming a member of its fire practitioner program:

- At a minimum, to be a fire practitioner, staff must complete standard coursework and pass a physical fitness capacity test. Annual "refreshers" and new fitness tests are required each year.
- To move from a basic fire practitioner into a more advanced position, such as an Engine Boss, staff must complete additional coursework and satisfactorily perform a series of tasks specific to the position while serving as trainees on burns.
- To become a qualified Burn Boss, staff might spend a decade or more completing all the coursework and experiential training that's required of this advanced position.

In 2023, the group agreed to a twenty-year partnership with the USDA Forest Service that uses both public and private funding to develop the prescribed fire workforce and increase collaboration. The 2023 agreement with the USFS called for a $45 million investment by the federal government over five years in collaborative, mutual priority prescribed fire projects. That includes twenty-one priority landscapes and 250 high-risk firesheds, primarily in the western United States, already identified through the Forest Service Wildfire Crisis Strategy. TNC has also worked at the legislative level. Prescribed burning is regulated at the state level, not the federal, and the organization's staff have lobbied to pass legislation and funding initiatives in states like California to address key issues that the group has prioritized. In 2008, TNC created TREX, or Prescribed Fire Training Exchange, to provide training opportunity to thousands of those in the fire profession. That includes the promotion of cooperative burning to help leverage each partner's skills and resources. Similarly, the Women-in-Fire-Training Exchanges, created in 2016, help women fire practitioners build support networks to advance their leadership.

Development of a more inclusive fire management workforce is key to TNC's mission. This involves fire practitioner job training and workforce development, with hiring that began in October 2023. One of the ways the workforce will grow involves the Forest Service Fuels Academy and the National Interagency Fire Training Center. Only about 15 percent of the federal fire management workforce is women, a number TNC plans to change. The organization is working with the USFS to expand the diversity of the agency's firefighting ranks through the Women in Fire Fellowship Program. The two-year program is planned to include four participants at a time, staggering each woman's starting year. The initial training includes everything from chainsaw certification to on-the-ground work in the Shawnee National Forest in southern Illinois.

Parallel to the workforce diversity initiative is TNC's North America Indigenous Landscapes and Communities Program, created in 2019, and the Indigenous Partnerships Program. Both emphasize the shift from a predominant Western science organization to one based on traditional knowledge by bringing indigenous populations to the table. In the Indigenous Partnerships Program, which has been working with four pueblos in northern New Mexico, leaders are developing a cultural fire plan for the Taos Pueblo to protect a tribal wilderness area in support of cultural traditions. Two other pueblos are creating a forest in the Jemez Mountains where 100,000 drought-resistant seedlings were planted in an area that burned from a fire in 2011. The hope is that the plants will adapt to intensifying climate change. TNC is also partnering with the Indigenous Peoples Burning Network to strengthen tribal capacity in the planning and implementation of cultural and traditional fire use.

The future of diversity programs was uncertain, however, with efforts by the second Trump administration to eliminate funding and support, including grants, for efforts focused on DEI—Diversity, Equity, and Inclusion—the organizational frameworks that seek to promote the full and fair treatment of all individuals.

Gifford Pinchot (1865–1946)

Gifford Pinchot has been the subject of numerous biographers who have written extensively about his career in forestry and politics, while his role in the development of the conservation movement, especially after the Second World War, has also been chronicled and lauded. Less well known is his perspective on forests and fires, which he referred to as "one of the most fruitful and fascinating of all the fascinating and fruitful, branches of forestry in the United States" (Pinchot, 1899).

Born in Simsbury, Connecticut, Pinchot was the son of a prominent New York City couple, James and Mary Pinchot, whose families included Union Army generals, a French immigrant, landowners, diplomats, and members of

Congress. He was schooled at home until he enrolled at the prestigious Phillips Exeter Academy, and then Yale University, from which he graduated in 1889. His graduation came at a time when forestry was just beginning to become a reputable and growing field of study in the United States. In order to broaden his knowledge, he attended the French National School of Forestry for a year to learn from a forestry curriculum that was more advanced than in his home country. In 1892, he became the manager of the forests belonging to the prominent Vanderbilt family at their large estate, Biltmore, near Asheville, North Carolina. He created the first large-scale forest management plan in the United States at a time when there were no professional foresters. He served at Biltmore for three years, then moved to New York to open a consulting business.

Pinchot had met the American naturalist, John Muir in 1893, and the two became friends. Pinchot toured the West with the National Forest Commission in 1896 as the only forester in the group, a trip that helped shape his beliefs on the role of forests that diverged from both Muir's views and those of the Commission. He advocated for a more utilitarian purpose for forests than Muir's preservationist perspective, and sought the formation of a professionally trained forest service. In 1897, he accepted a position with the US Department of the Interior, and a year later, became the head of the Division of Forestry within the Department of Agriculture. Using the concept of planned use, Pinchot's reforms of the organization called for scientific forestry and a culture supportive of professionalization. In keeping with that ethic, his family endowed the Forest School at the Yale School of the Environment. When the US Forest Service was created in 1905 and the forest reserves were transferred to the Department of Agriculture, he became its first chief, serving until 1910.

In 1899, Pinchot published a notable article in the magazine, *National Geographic*, "The Relation of the Forests and Forest Fires." As fire ecologist James Agee (2011) notes, Pinchot was, at the time, a forester without portfolio, since the forest reserves were managed by the Office of Public Lands within the Department of the Interior, while Pinchot was part of the Department of Agriculture's Bureau of Forestry. Pinchot did, however, recognize the study of forest fires as an emerging area of importance that had not yet been investigated. On one hand, he refers to the "creative action of forest fires" while on the other, he believes there is a need to find out "how and why fires do harm, and how best they may be prevented or extinguished" (Pinchot, 1899). Like many of his contemporaries, fires were considered in terms of the dollar cost they caused, although he admitted there is other damage to be considered, such as the impact on water supplies, the deterioration of the soil, and the destruction of young growth. Perhaps based on his study of forests in Europe and his trip to the Pacific Northwest, Pinchot notes that the composition of a forest may be based on the resistance to wildfires as a quality which some types of trees possess. Rather than admitting that fire has a beneficial role, however, Agee notes that Pinchot concludes by writing, "I hasten to add that these facts do not imply any desirability in the fires which are now devastating the West" (Pinchot, 1899).

Had Pinchot done so, the fire policies that would later govern the US Forest Service might have undergone a totally different trajectory. Instead, once the forest reserves were transferred from the Department of the Interior to the Department of Agriculture, Pinchot embarked on the creation of management guidelines for the agency that became known as the *Use Book*, published in 1905. He underscored the USFS duty, as he put it, to protect the reserves from forest fires, calling that duty insurance against the destruction of property, timber resources, and water supply (Pinchot, 1905). His perspective was underscored in 1910 by the Great Fire that made exclusion and suppression the agency's mission for decades to come.

Pinchot's political career represents a shift in his life that began after President Taft dismissed Pinchot as Chief in January 1910. Pinchot and his allies started a new group, the Progressive Party, that supported fellow conservationist Theodore Roosevelt for president in 1912. The two men had similar views on the place of forests and natural areas that would influence many of Roosevelt's policies. When Roosevelt lost the election, Pinchot continued his work with the Progressives, running unsuccessfully for the US Senate from Pennsylvania in 1914. He was appointed as chair of the Pennsylvania Forest Commission in 1920, and replaced the group's political appointees with professional foresters. Two years later, he ran successfully for governor of Pennsylvania in 1922, not because of his credentials in forestry but because he supported Prohibition. He ran again for governor in 1930, working to help indigent and impoverished citizens, supporting the Civilian Conservation Corps' projects on Pennsylvania's public lands, expanding state parks, and paving rural roads. He ran unsuccessfully for the US Senate in 1934, and was unsuccessful in running again for governor in 1938. Pinchot continued to write about forestry and on his memoirs, which were published after his death.

Edward Crockett Pulaski (1866–1931)

The story of the man who has been called the hero of the conflagration that destroyed about a third of the town of Wallace, Idaho, where he saved all but a handful of his firefighting crew, is almost mythical. There is a hero, a Great Fire, and great drama that is an important part of American firefighting history.

"Big Ed" Pulaski, as he is known, incorrectly thought to be the great-grandson of a Polish count, was the son of Polish immigrant Rudolph Pulaski and his wife Celia. The couple married in 1864 in Tiffin, Ohio, and Ed was born nearby in Green Springs, where he attended a country school. Sometime in the late 1870s the family moved to Marshall in central Missouri, where Ed continued his education until age fifteen. He was inspired to travel by his uncle, Edwin Crockett, who had found financial success mining gold in Idaho after traveling to California and British Columbia. After listening to his uncle's stories of western adventure and his service in the Civil War, Ed left home in 1884, traveling on the Northern Pacific Railway to the Montana territory. From there, he trekked to the booming town of Murray in the Idaho Territory, where he found work as a packer. Pulaski had numerous jobs after leaving Missouri, working in sawmills, mines, and on a cattle ranch, finding work as a blacksmith, steam engineer, and builder along the way. He married Emma Zenobia Dickinson in 1900, and together they adopted a daughter, Elsie.

They moved to Wallace, Idaho, a small mining town that served as the county seat for Shoshone County with a census population of about 900 in 1890 and 3,000 by 1910. The town had been founded in 1884 after gold had been discovered in a stream nearby in 1882. Wallace was named for Civil War Veteran William R. Wallace, whose Wallace Townsite Company sold lots in the new settlement, seeing potential in the area's gold (and later, silver) mining. In 1890, strong winds fed a fire that destroyed much of the business center, leading the town to organize a better-equipped fire company and passage of an ordinance requiring fireproof construction downtown.

Pulaski joined what was then called the National Forest Service as an assistant ranger in 1908. He had no training in forestry, and less education than his colleagues, but he was familiar with the Coeur d'Alene National Forest and the region because of the time he spent prospecting there. The spring of 1910 was ominously dry throughout Idaho, Washington, and Montana. In the white pines of the Coeur d'Alene National Forest, firefighters had been battling fires since early April due to extremely dry conditions in the area. Because of a spring/summer drought, electrical storms ignited even more blazes across the Northern Rockies. But there were also fires started by the trains that burned coal and residents who had, he believed, no respect for forest fire laws and those who had set fires deliberately for various purposes. In mid-August, President William Howard Taft sent soldiers from the 25th Infantry (known as the Buffalo Soldiers) from Spokane, Washington, to help defend the town of Wallace from fires in the surrounding mountains. On the evening of August 20, 1910, warm hurricane-strength winds, called a Palouser, a Sahaptin Indiana word, whipped through the town. Embers ignited trash adjacent to the local newspaper's building, and flames then spread to nearby offices as local residents rushed to board trains to evacuate women and children to other towns. Wallace had only a one-wagon fire department, and Pulaski and a crew of forty-five men were fighting fires on a ridge about 3 miles out of town while his wife and daughter stayed in Wallace until a neighbor banged on their door and told them they had to leave. Pulaski was in charge of a force of about 200 men, many of them immigrants with no experience in fighting fires who worked long days with little sleep. They had to be taught even the basics—staying hydrated, how to hold an ax, and some just walked away from the job where most had never received a promised paycheck. Pulaski divided them into two groups working the mountain ridge, with most assigned to keep the fires from spreading into the town of Wallace.

Initially, Pulaski told his men that they needed to head down the West Fork of Placer Creek toward safety in Wallace, but after running into flames headed their way, he revised course to get them to the War Eagle Mine. When that route became blocked by flames, he directed them to the abandoned Nicholson Mine for shelter, fighting the fire at the mouth of the tunnel. All but one of his men and two horses made it there safely, although all suffered burns and smoke inhalation. Inside the mine, five more men and the two horses died, but the other thirty-eight survived after Pulaski led them back to the hospital in Wallace. Pulaski spent months in the hospital, later returning to work blinded in one eye and with damaged lungs from the fires.

A year after the fires, called The Big Burn, subsided, Pulaski was credited with making a prototype of a firefighting hand tool he gave his own name, the "pulaski"—a type of ax and hoe combined into a single implement that the Forest Service began using in 1913. The Pulaski was widely used to make firebreaks, and was commercially manufactured as early as 1920. While other men and the Collins Tool Company have been said to have made a similar tool before him, the name stuck. He left the Forest Service in 1929, and died two years later in Coeur d'Alene from complications from injuries he received in an automobile accident. A nearby peak, Mt. Pulaski, was later named in his honor, and local residents raised $45,000 for a granite memorial in Wallace similar to what Pulaski had designed.

Pulaski has sometimes been called a hero, not just for his efforts during the fires, but also because he sought to have the hospital costs incurred by his men covered by the government, and because he wanted some sort of recognition for the men who had died in the fire. Pulaski never received any compensation for himself for the cost of his medical care. In 1921, Congress appropriated $500 to the Coeur d/Alene National Forest for a suitable monument, located in the Nine Mile Cemetery outside Wallace.

Stephen J. Pyne (1949–)

One of the most prolific authors on the subjects of fire, Antarctica, and the Grand Canyon, winner of the prestigious MacArthur Fellowship, now Professor Emeritus at Arizona State University, Stephen J. Payne admits that he has not always been successful. Growing up in a suburb of Phoenix, Arizona, as a boy he spent time reading about history, and at age twelve, he developed strep infections that left him an invalid. He attended a Jesuit high school (graduating as valedictorian), and at eighteen, he joined the North Rim Longshots fire crew at the Grand Canyon—a job he would stay with for fifteen seasons as saw boss and then crew boss. During those same years, he attended Stanford University from 1967 to 1971 where he received his Bachelor of Arts degree in English. Despite being rejected by every graduate program he applied to but one, he attended the University of Texas, Austin, from 1972 to 1976, receiving his master's and PhD degrees in American Civilization.

Admitting he lived two separate lives, he worked fires during the summer, becoming a researcher in a field he helped create—fire history—in the winter. He had a third interest as well, marrying his wife Sonja on the rim of the Grand Canyon, and later, becoming a father. Pyne had one resume built on his wildland fire expertise and experience with Grand Canyon National Park, Rocky Mountain National Park, and Yellowstone National Park during the late 1970s and 1980s, along with professional service to academic organizations and as a consultant. A different path found him starting his academic career with a history position at the University of Iowa from 1981 to 1986; he returned to his home turf in 1985, teaching and researching at Arizona State University until his retirement in 2018.

Pyne readily admits that his tenure at Grand Canyon National Park gave him what he calls "woods credibility" because they helped shape how he now thinks and speaks about fire. As a member of a fire crew, he could write about the language of fire and build a relationship with his subject in a way few other scholars have done. Like his dual career paths of wildfire and academia, Pyne has managed to balance his writing with the publication of more than forty books, beginning with an expansion of his dissertation on American geologist, Grove Karl Gilbert in 1980. His first wildfire book, *Fire in America*, was published in 1982 as what he called a cultural history, while he remained, as he put it, "a smokechaser at heart, hotspotting from one topic to the next." His next major writing foray took him to the end of the earth and his 1986 book, *The Ice: A Journey to Antarctica*, and a series of additional histories of fire in Canada, Australia, Europe, and major fires in the United States. Although his publications about wildfire cover numerous themes, he occasionally has returned to his second home, the North Rim of the Grand Canyon, with *Dutton's Point: A Natural History of the Grand Canyon* (1983); *Fire on the Rim: A Firefighter's Season at the Grand Canyon* (1989); and *How the Canyon Became Grand: A Short History* (1998). In 2009, he also wrote a primer of sorts based on his own experience, *Voice and Vision: A Guide to Writing History and Other Serious Nonfiction*, and in 2018, *Style and Story: Literary Methods for Writing Nonfiction.*

One of the factors that led Pyne to be considered as one of the nation's leading environmental historians is that he has also done comparative research, including fire histories on Australia, Canada, Europe, and Russia. More recently, he used his experiences as a commentator on the 2025 Los Angeles wildfires. Five years before the Palisades and Eaton Fires, he had been interviewed for the History News Network where he commented, "California is built to burn and to burn explosively. Against that, we have a society that is determined to live and work where and how it wants." These seemingly prescient words were also placed in comparative context when he noted, "Australia is California at a continental scale, only drier, and with winds that can turn the southeast into a veritable fire flume" (Crawford, 2020).

The list of Pyne's awards and recognition is also bifurcated between his careers, starting in 1973, when he received an Outstanding Performance

Award from the National Park Service. From then on, he received awards from the major professional history and science societies, to teaching awards and as a faculty member, to fellowships and as an invited lecturer. In 1992 he was awarded a Fulbright Fellowship to conduct research in Sweden. One of the most prestigious awards was bestowed upon Pyne in 1988 when he was only thirty-nine years old—the MacArthur Foundation Fellowship.

Throughout his career, Pyne developed his expertise as the go-to person when reporters as well as officials sought out his opinions and perspectives on fire. He has participated in countless workshops, on panels, and as a member of advisory groups from Yellowstone to Ghana to Australia to Maine. In these capacities, Pyne has been a rare type of scholar—the kind who not only publishes his work but also teaches and consults with key wildfire agencies and organizations throughout the globe.

Before he retired, he and his eldest daughter, Lydia, wrote a book about the Pleistocene era entitled *The Last Lost World* (2012), and in 2021, a book that received considerable interest, *The Pyrocene: How We Created an Age of Fire and What Happens* Next that provides a new perspective on his previously published work. Pyne said he retired from his academic career when his oldest grandchild was the same age as his youngest students. He now lives near Phoenix and calls himself a writer and urban farmer, tending Tunis sheep, chickens, vegetables, and citrus trees. As he said at his retirement luncheon, "I don't intent to shut down my mind. I won't quit writing. I'm just through teaching. It's time" (Pyne, 2018).

Jan Willem van Wagtendonk (1940–2022)

In a profile interview published by the US Geological Survey in 2022, the author began by noting "The first thing to know about emeritus scientist Jan van Wagtendonk is that he loves trees—always has always will" (Western Ecological Research Center, 2022), Three years later, van Wagtendonk passed away at his home in Winters, California, not far from the Mariposa Grove of sequoias where the Washburn Fire was burning. The grove and its famous trees survived the fire largely because van Wagtendonk had helped to develop the first fire prescriptions for Yosemite during his career with both the USGS and the US Forest Service.

His parents, Willem Johan van Wagtendonk and Roelina Jantina Lottering, emigrated to the United States from Rotterdam in The Netherlands in 1939, and he was born in Palo Alto, California. As an impressionistic teenager on a two-month-long camping trip with his parents and older sister, he met a forest ranger and decided that he, too, could work with trees for a living. He moved with his family to Bloomington, Indiana, in 1946 and attended school there. He was admitted to Purdue University's forestry school, and found a summer job on a wildland firefighting crew in Oregon. The experience was transformative, and he decided to transfer from Purdue to attend Oregon State University while working on summer fire crews in Oregon and then Alaska. One of his jobs from 1960 to 1961 was as a smoke jumper, and he was also a smoke jumper in Fairbanks, Alaska.

He graduated with a bachelor's degree in forestry from Oregon State in 1963, and then joined the US Army as an officer in the 101st Airborne Division and as an advisor to the Vietnamese army. In 1967 he applied to graduate school at the University of California, Berkeley, receiving a master's degree in range management in 1968 and his PhD in wildland resource science with a specialty in fire ecology in 1972. His mentor at Berkeley, Dr. Harold Biswell, played a pivotal role in shifting a century of National Park Service fire suppression policy. Dr. Biswell conducted studies in the 1960s and 1970s at Whitaker's Forest just outside of Kings Canyon National Park that demonstrated the vital ecological role that fire played within giant sequoia groves. Following his observations, Kings Canyon and Sequoia instituted a new fire policy of allowing natural fires to burn and conducted prescribed burns in areas long subject to fire suppression. Interest in Biswell's work extended to Yosemite National Park, and Robert Barbee, the natural resources manager for the park, consulted with Biswell about how fire might be reintroduced into Yosemite.

In van Wagtendonk's profile interview, he mentioned that in forestry school, fire ecology was only obliquely mentioned, and fire was certainly not something you tell people you'll set. Dr. Biswell's controversial research on prescribed burning was not yet accepted within the discipline. Jan's doctoral dissertation was focused on refining burning prescriptions for backing and head fires in mixed-conifer forests, and his experiments were designed to develop quantitative burning prescriptions based on Dr. Biswell's work. The US Forest Service, however, refused to give him permission to burn on their lands. Instead, he accepted an invitation from the National Park Service, which had begun incorporating prescribed fire programs in the late 1960s. Van Wagtendonk went to Yosemite National Park to conduct his dissertation research, and subsequently was offered a research position with the park. He later held a similar position with the USGS and served as one of the original scientists at the Western Ecological Research Center before returning to the NPS, where he retired as emeritus scientist.

During his career, van Wagtendonk's research agenda focused on two subjects: fire science and recreation management, most of which he conducted at Yosemite. Much of his work was conducted on Yosemite's Illilouette Creek Basin in the early 1970s. He informally served as a fire historian, chronicling the role of fire in the park and how it has been managed. Later, he added climate change to his study of fire behavior, noting that larger fires burned in California as the climate warms. His work incorporated new technology to the study of fire, as well as the influence of fire on small mammals and birds like owls. The observations he made during his dissertation research are part of his legacy, forming the basis of today's existing Yosemite fire management plan. As he examined Yosemite's backcountry wilderness areas, he found that the park was being loved to death by visitors and campers who had trampled natural areas and overflowed the park's most popular camping sites. His observations and published research led to mandatory backcountry permit and quota systems at a time when visitation was skyrocketing in the park.

One memorial essay notes that van Wagtendonk was known for always giving of his time to talk to other about fire and forests, mentoring graduate students and inviting him into his home and professional life. He led field courses for the California Association for Field Ecology and led annual backpacking trips to Yosemite, helped to found the Association for Fire Ecology, and became its president and editor of the association's journal. During his career, he authored more than 140 papers, chapters in books, and technical reports for agencies. He won numerous awards for his work from the National Park Service, Department of the Interior, and fittingly, the Biswell Lifetime Achievement Award, named after his university graduate mentor.

Further Reading

Agee, James K. 1995. "Memorial Dedication to Dr. Harold H. Biswell." USDA Forest Service. General Technical Report. PSW-GTR-158. https://doi.org/10.2737/PSW-GTR-158.

Agee, James K. 2011. "Introduction to Gifford Pinchot's Article." *Fire Ecology* 7 (2). https://fireecology.springeropen.com/articles/10.4996/fireecology.0703002.

Agee, Jim, Scott Stephens, Andi Thode, and Allie Weill. 2022. "In Memoriam: Jan van Wagtendonk 1940-2022." *Association of Fire Ecology*. August 11. Fireecology.org.org/news/wagtendonk?srsltid.

AllGov. "Abigail Kimbell." www.allgov.com/officials/kimbell-abigail?officialid=28556.

Arno, Stephen. 2014. "Slow Awakening: Ecology's Role in Shaping Forest Fire Policy." *Forest History Society*. October 16. https://fhsarchives.wordpress.com/2014/10/166295_arno_role_of_ecology_in_shaping_forest_fire_policy/.

Associated Press. 1988. "Yellowstone Superintendent Is Taking the Heat for Fire Policy." *Deseret News*. September 25. https://www.deseret.com/1988/9/25/18779055/yellowstone-superintendent-is-taking-the-heat-for-fire-policy/.

Associated Press. 2016. "Head of Yellowstone during 1988 Fires Dies." *Wyoming News Now*. October 4. https://www.wyomingnewsnow.tv/archive/head-of-yellowstone-during-1988-fires-dies/article_a4467746-6e01-50c1-86b8-822c2fbcee33.html.

Benallie, Kalle. 2023. "Indigenous Leadership in Nature Conservancy Hope to Build Tribal Support." *ICT News*. November 9. https://ictnews.org/news/indigenous-leadership-in-nature-conservancy-hope-to-build-tribal-support.

Biswell, Harold H. 1989. *Prescribed Burning in California Wildlands Vegetation Management*. Berkeley: University of California Press.

Congressional Record. 2009. "Commending Abigail Kimbell." Vol. 155. Part 14. https://www.govinfo.gov/content/pkg/CRECB-2009-pt14/html/CRECB-2009-pt14-Pg18434-3.htm.

Crawford, Matthew. 2020. "A History of Fire: An Interview with Stephen Pyne." *History News Network*. January 12. https://www.historynewsnetwork.org/article/the-history-of-fire-an-interview-with-stephen-pyne.

Danielson, Stentor. 2021. "Stephen J. Pyne, The Pyrocene: How We Created an Age of Fire, and What Happens Next." *New Books in Geography*. September 8. https://www.environmentandsociety.org/mml/interview-stephen-j-pyne-author-pyrocene-how-we-created-age-fire-and-what-happens-next.

Department of Agriculture, Forest Service (Completion Report). *The Gisborne Era of Forest Fire Research*. Missoula: University of Montana Forest and Conservation Experiment Station in cooperation with the U.S. Depart of Agriculture, 1977.

Egan, Timothy. 2009. *The Big Burn: Teddy Roosevelt and the Fire That Saved America*. Boston, MA: Houghton Mifflin Harcourt.

Ellison, Jon. 2019. "Into the Woods: Through Thick and Thin, the Biltmore Forest School Planted the Seeds of Modern Forestry." *WNC Magazine*. July–August. https://wncmagazine.com/feature/woods.

Forest History Society. n.d.a. "Abigail Kimbell: 16th Chief of the Forest Service, 2007–2009." https://foresthistory.org/research-explore/us-forest-service-history/people/chiefs/abigail-kimbell/.

Forest History Society. n.d.b. "Harry T. Gisborne." https://foresthistory.org/research-explore/us-forest-service-history/people/scientists/harry-t-gisborne/.

Forest History Society. n.d.c. "William B. Greeley." https://foresthistory.org/research-explore/us-forest-service-history/people/chiefs/william-b-greeley-1879-1955/.

Gabbert, Bill. 2019. "The True Story of the Pulaski Fire Tool." *Wildfire Today*. August 19. https://wildfiretoday.com/2019/08/19/the-true-story-of-the-pulaski-fire-tool.

Greeley, William B. 1953. *Forest Policy*. New York: McGraw-Hill.

Hannibal, Mary Ellen. 2021. "The History and Future of Fire." *Science*. October 14. https://www.science.org/doi/10.1126/science.abl9129.

"Jan Willem van Wagtendonk: Memorial." *Find a Grave*. https://www.findagrave.com/memorial/242656361/jan-willem-van_wagtendonk.

Jones, Lyndsay. 2024. "A Bloomington Native Is Working to Bolster—and Diversify—the Ranks of Wildland Firefighters." WGLT Public Radio. August 21. https://www.wglt.org/local-news/2024-08-21/a-bloomington-native-is-working-to-bolster-and-diversify-the-ranks-of-wildland-firefighters.

Keane, Bob. 2022. "A Tribute to Steve Arno, Pioneering Forester and Friend." August 23. *Whitebark Pine Ecosystem Foundation*. https://whitebarkfound.org/tribute-to-steve-arno/.

Komarek Family Papers. Chicago Academy of Sciences. Peggy Notebaert Nature Museum, Chicago, Illinois.

Kramer, Becky. 2010. "Heyburn Left Thorny Legacy on Natural Resources." *The Spokesman-Review*. August 22. https://www.spokesman.com/stories/2010/aug/22/heyburn-left-thorny-legacy-natural-resources/.

Landers, J. Larry. "About E.V. Komarek." https://talltimbers.org/wp-content/uploads/2018/09/3-Landers1991a_op.pdf.

Lawatch, Brian. n.d. "Interview with Sue Husari." Wildland Fire Leadership Development Program. https://fs-prod-nwcg.s3.us-gov-west-1.amazonaws.com/s3fs-public/2023-06/sue-husari.pdf?VersionId=x90XlOGdXHjlH7qL2xFZ1ZVJb7g Lewry.

Lunder, Zeke. 2021. "A Conversation with Sue Husari." *The Lookout*. November 21. https://the-lookout.org/2021/11/21/a-conversation-with-sue-husari/.

Maclean, Norman. 1992. *Young Men and Fire*. Chicago, IL: University of Chicago Press.

Miller, Char. 2001. *Gifford Pinchot and the Making of Modern Environmentalism*. Washington, DC: Island Press.

Monthei, Amanda. 2020. "The First Women on Hotshot Crew." *Outside Online*. July 23. https://www.outsideonline.com/outdoor-adventure/exploration-survival/first-women-hotshots-wildland-firefighters/.

Morgan, George T., Jr. 1961. *William B. Greeley: A Practical Forester.* St. Paul, MN: Forest History Society.

National Park Service. n.d. "The Yellowstone Fires of 1988." https://www.nps.gov/yell/learn/nature/1988-fires.htm.

The Nature Conservancy. 2025. "Why We Work with Fire." January 21. https://www.nature.org/en-us/what-we-do/our-priorities/protect-water-and-land/land-and-water-stories/why-we-work-with-fire./.

Osterink, Carole. 2014. "A Life of Achievement Marred by Tragedy." *The Gossips of Rivertown*. October 31.

Paskus, Laura. 2008. "Up in Smoke: Obama Administration Will Inherit a Beleaguered Forest Service." *High Country News*. December 18. https://www.hcn.org/issues/40-23/up-in-smoke/.

Pinchot, Gifford. 1899. "The Relation of Forests and Forest Fires." *National Geographic* X: 393–403.

Pinchot, Gifford. 1905. *The Use of the National Forest Reserves*. Washington, DC: USDA Forest Service.

Pinchot, Gifford. 1947. *Breaking New Ground*. New York: Harcourt Brace Jovanovich.

Pyne, Sonia. n.d. "Edward Crockett Pulaski: A Short Biography." https://www.fs.usda.gov/Internet/FSE_DOCUMENTS/stelprdb5442832.pdf.

Pyne, Stephen J. 2008. *Year of the Fires: The Story of the Great Fires of 1910*. Missoula, MT: Mountain Press Publishing.

Pyne, Stephen J. 2018. "Scholar on Fire, Soul on Ice. What I've Learned." November. https://www.stephenpyne.com/bio.html/.

Schenck, Carl Alwin, and Ovid Butler. 1998. *Cradle of Forestry in America: The Biltmore School of Forestry 1898–1913*. Durham, NC: Forest History Society.

"Senator Objects to Appointment of New Forest Service Chief." 2007. *Washington Post*. January 13. https://www.washingtonpost.com/wp/dyn/content/article/2007/01/12/AR2007011201974.html.

Southard, Lewis F. 2011. "The History of Cooperative Forest Fire Control." *Forest History Today*. Spring/Fall. https://foresthistory.org/wp-content/uploads/2017/01/Cooperative_Fire_Control.pdf.

"Stephen Arno Obituary." Legacy.com. https://www.legacy.com/us/obituaries/missoulian/name/stephen-arno-obituary?id+35172653.

"U.S. District Forester Resigning—Colonel Coert DuBois to Enter Consular Service." 1919. *Sausalito News* 35 (43). https://cdnc.ucr.edu/?a=d&d=SN19191025.2.55.

USDA Forest Service. 1910. Proceedings of the Supervisor's Meeting, District 5. San Francisco, California, December 13–19. Box 1: Historical Manuscripts, 1902-1913, Record Group 95, National Archive Record Center, San Bruno, California.

Van Wagtendonk, Jan W. 1995. "Dr. Biswell's Influence on the Development of Prescribed Burning in California." USDA Forest Service. General Technical Report. PSW-GTR-158. https://doi.org/10.2737/PSW-GTR-158.

Waldorf, G. B. 1991. "History of Birdsong." https://birdsongnaturecenter.org/page3/page3.html.

Western Ecological Research Center. 2022. "Remembering Jan van Wagtendonk, Who Shaped Fire and Recreation Management in the Iconic Yosemite National Park." U.S. Geological Survey. https://www.usgs.gov/news/remembering-jan-van-wagtendonk-who-shaped-fire-and-recreation-management-iconic-yosemite.

World Forestry Center. 2016. "William B. Greeley 1879-1955." *Evergreen Magazine*. https://evergreenmagazine.com/content/files/2025/03/GREELEY-WILLIAM-B.pdf.

5

Data and Documents

Data

The following four tables illustrate some of the many ways wildfires can be measured and described, from the value of the losses seen in Table 5.1 to lives lost, the amount of land burned, and the number of reported wildfires. Several trends emerge from the numbers. One, the costliest wildfires, with the exception of the Tunnel Fire in 1991, have all occurred since 2017. Despite all the improvements in firefighting equipment, training, and management, wildfire costs have skyrocketed, even without the addition of the Palisades and Altadena Fires of 2025, for which the total costs are still being calculated. Two, in contrast, the number of persons who have died in wildfires, seen in Table 5.2, has gone down since the turn of the twentieth century, in part due to changes in building codes and materials, and the creation of wildland firefighting units with better trained personnel to minimize the loss of life. Tables 5.3 and 5.4 indicate that the number and scale of wildfires each year is variable. In recent years, more wildfires have occurred in the East, but those in the West have been larger, burning more acreage and often in undeveloped areas in states like Alaska and Texas. Lastly, it is important to note that statistics alone, such as acreage burned or the number of wildfires in a year, fail to indicate the overall impact of a wildfire which may include the ecological effects, long-term impact on a community's economy, or the health of its citizens.

A fire investigator from the Bureau of Alcohol, Tobacco, Firearms and Explosives looks for evidence that might point to the source of the deadly wildfire that raged through Lahaina, Maui, Hawaii, August 18, 2023. More than a thousand people filed suits against seven defendants accused of causing the tragedy, resulting in a $4.04 billion settlement agreement in 2024 (US Customs and Border Protection photo by Glenn Fawcett).

Table 5.1 Top Ten Costliest US Wildfires, by Dollar Loss 2024 Value (1993–2024)

Fire	Month/Year	Dollar Loss (2024 value)
Camp	11/2018	$12,400,000,000
Tubbs	10/2017	11,049,000,000
Lahaina	8/2023	6,180,000,000
Woolsey	11/2018	5,208,000,000
Tunnel/Oakland Hills	10/1991	3,893,000,000
Atlas	10/2017	3,810,000,000
Glass	9/2020	3,540,000,000
Thomas	12/2017	3,175,000,000
CZU Lightning Complex	10/2020	3,000,000,000
LNU Lightning Complex	8/2020	2,700,000,000

Source: National Fire Protection Association. https://www/nfpa.org/education-and-research/research/nfpa-research/fire-statistical-reports/wildland-fire-statistics. Reproduced with permission of NFPA from NFPA Research Archive Files. Copyright © 2024, National Fire Protection Association. For additional information on the subject, please go to www.nfpa.org.

Table 5.2 Top Ten Deadliest US Wildfires, by Human Lives Lost (1871–2024)

Fire	Month/Year	Fatalities
Peshtigo	10/1871	1,152
Cloquet and Moose Lake	10/1918	453
Great Hinckley	9/1874	418
Thumb	9/1881	282
Lahaina	8/2023	101
Great	9/1910	87
Camp	11/2018	85
Yacolt Burn	9/1902	65
October Fire Siege	10/2017	44
2020 Fire Siege	8/2020	31

Source: National Fire Protection Association. https://www.nfpa.org/education-and-research/research/nfpa-research/fire-statistical-reports/wildland-fire-statistics. Reproduced with permission of NFPA from NFPA Research Archive Files. Copyright © 2024, National Fire Protection Association. For additional information on the subject, please go to www.nfpa.org.

Table 5.3 Top Ten Largest US Wildfires, by Acres Burned (1898–2024)

Fire	Month/Year	State	Acres Burned
Great Fire of 1910	8/1910	Idaho/Montana	3,000,000
Great Fire of 1898	2/1898	South/North Carolina	3,000,000
Taylor Highway Complex	6/2004	Alaska	1,303,358
Ruby	6/1940	Alaska	1,240,000
Peshtigo	10/1871	Wisconsin	1,200,000
Kateel River #5	6/1957	Alaska	1,161,200
Smokehouse Creek	2/2024	Texas	1,054,000
August Complex	8/2020	California	1,032,648
Thumb	9/1881	Michigan	1,000,000
Unalakleet	1941	Alaska	1,000,000

Source: National Fire Protection Association. https://www.nfpa.org/education-and-research/research/nfpa-research/fire-statistical-reports/wildland-fire-statistics. Reproduced with permission of NFPA from NFPA Research Archive Files. For additional information on the subject, please go to www.nfpa.org.

Table 5.4 Total US Wildland Fires and Acres Burned, by Year (1984–2024)

Year	# Fires	Acres Burned
1984	20,493	1,148,409
1985	82,591	2,896,147
1986	85,591	2,719,162
1987	71,300	2,447,296
1988	72,750	5,009,290
1989	48,949	1,827,310
1990	66,481	4,621,621
1991	75,754	2,953,578
1992	87,394	2,069,929
1993	58,810	1,797,574
1994	79,107	4,073,579
1995	82,234	1,840,546
1996	96,363	6,065,998
1997	66,196	2,856,959
1998	81,043	1,329,704
1999	92,487	5,626,093

Table 5.4 (Continued)

Year	# Fires	Acres Burned
2000	92,250	7,393,493
2001	84,079	3,570,911
2002	73,457	7,184,712
2003	63,629	3,960,842
2004	65,461	8,097,880*
2005	66,753	8,689,389
2006	96,385	9,873,745
2007	85,705	9,328,045
2008	78,979	5,292,468
2009	78,792	5,921,786
2010	71,971	3,422,724
2011	74,126	8,711,367
2012	67,774	9,326,238
2013	47,579	4,319,546
2014	63,312	3,595,613
2015	68,151	10,125,149
2016	67,743	5,509,995
2017	71,499	10,026,086
2018	58,083	8,767,492
2019	50,477	4,664,364
2020	58,950	10,122,336
2021	58,985	7,125,643
2022	68,988	7,577,183
2023	56,580	2,693,910
2024	64,897	8,924,884

*Figures for 2004 do not include state lands for North Carolina.
Source: National Interagency Coordination Center. nifc.gov/fire-information/statistics/wildfires.

Documents

"The Relation of Forests and Forest Fires," Gifford Pinchot (1899)

It is sometimes difficult to remember that our knowledge and concern about wildfires could be said to be a relatively modern phenomenon. In this excerpt from the popular publication National Geographic *just before the turn of the century,*

Gifford Pinchot, who held the position of US Forester within the US Department of Agriculture, reminds us that the nation was just beginning to equate forests and fire with the timber industry, the scientific study of trees, and the damage wildfires cause. The $20 million cost attributed to wildfires covers the loss of timber, and nothing else at this time. Only with additional study, Pinchot writes, will we begin to know the additional losses that fire can produce. The article comes before the Great Fires of 1910 will demonstrate the other types of economic and human costs that ensue as our perspectives change.

The study of forest fires as modifiers of the composition and mode of life of the forest is as yet in earliest stages. Remarkably little attention, in view of the importance of the subject, has hitherto been accorded to it. A few observers who have lived much with the forest, such as John Muir of California, have grouped fire with temperature and moisture as one of the great factors which govern the distribution and character of forest growth; but so little has been said or written upon the subject that the opinion of each man seems to have been reached independently and upon the single basis of personal observation. The documents upon the subject still reside, with very few exceptions, in the forest itself. It is unfortunate that our acquaintance with what might almost be called the creative action of forest fires should be so meager, for only through a knowledge of this relation and through the insight which such knowledge brings can there be gained a clear and full conception of how and why fires do harm, and how best they may be prevented or extinguished. The records of past fires, written in the forest now on the ground, are often decipherable for more than a hundred years back, and in many cases for more than a hundred years back, and in many cases for more than twice that length of time. Such records throw light on the relations of forests and fires as nothing else can, and are consequently the most valuable of all documents upon the somewhat intricate but most important question of the final effect of fire upon the forest; for we must clearly realize, before the present subject can fall into its proper sequence, that we have not stated everything when we say that "a given forest is destroyed by fire."

That fires do vast harm we know already, although just what the destruction of its forests will cost the nation is still unknown. Records compiled by the Division of Forestry indicate that the average direct recorded loss from this source is not less than $20,000,000 a year. To this figure must be added that vast direct loss unrecorded, together with a great but indefinite damage from the effect of forest destruction on water supply, and other losses of immense importance, the deterioration of the soil, the destruction of the young growth, and the loss of the increment which a healthy young forest would have been laying on year by year. The latter may commonly amount in a pure forest to several hundred board feet per acre each year. With further study a more exact statement of the grand total of the loss will be possible; but even now it is safe to assume that for the nation as a whole the loss is represented yearly by a sum much in excess of $50,000,000. That figure sufficiently proves that the destructive action of fire on the forest in its relation to human needs is a subject of the first interest and importance.

Source: Pinchot, Gifford. 1899. "The Relation of Forests and Forest Fires." *National Geographic* X. https://foresthistory.org/wp-content/uploads/2017/01/pinchot.pdf.

Weeks Act (1911)

The disastrous 1910 fire season, and the need to protect valuable timber resources, served as the impetus for the federal government's modern efforts to locate, purchase, and protect private land as national forests. An estimated 3 million acres had burned in Montana, Idaho, and Washington; 8 billion board feet of timber were destroyed; and eighty-five people, including seventy-two firefighters, perished in the blazes, which became known as The Big Burn in subsequent accounts. Introduced into Congress by John W. Weeks of Massachusetts, the statute appropriated $9 million for the purchase 6 million acres of land in the eastern United States. The statute, which was signed into law by President William Howard Taft, also authorized the Secretary of Agriculture to cooperate with the states in the organization and maintenance of a fire protection system to protect private and state lands, important because 80 percent of all forested land in the country was privately owned and unprotected against wildfire, and the concept of a federal-state partnership was new and untried. Congress set aside $200,000 in matching funds for states with forest protection agencies—only sixteen of the forty-six states had them at the time—but there was little left to actually fight fires. While important in helping to encourage cooperation and institutional support, the restrictive nature of the Weeks Act would be amended in 1924 with the passage of the Clarke-McNary Act. An excerpt from the Weeks Act is reproduced here.

ACT OF MARCH 1, 1911-(WEEKS LAW) [As Amended Through P.L. 110–343, Enacted October 3, 2008] CHAP. 186.—An Act To enable any State to cooperate with any other State or States, or with the United States, for the protection of the watersheds of navigable streams, and to appoint a commission for the acquisition of lands for the purpose of conserving the navigability of navigable rivers. Be it enacted by the Senate and House of Representatives of the United States of America in Congress assembled, That 16 U.S.C. 552 the consent of the Congress of the United States is hereby given to each of the several States of the Union to enter into any agreement or compact, not in conflict with any law of the United States, with any other State or States for the purpose of conserving the forests and the water supply of the States entering into such agreement or compact. SEC. 2. 16 U.S.C. 563. That the sum of two hundred thousand dollars is hereby appropriated and made available until expended, out of any moneys in the National Treasury not otherwise appropriated, to enable the Secretary of Agriculture to cooperate with any State or group of States, when requested to do so, in the protection from fire of the forested watersheds of navigable streams; and the Secretary of Agriculture is hereby authorized, and on such conditions as he deems wise, to stipulate and agree with any State or group of States to cooperate in the organization and maintenance of a system of fire protection on any private or state forest lands within such State or States and situated upon the watershed of a navigable river: Provided, That no such stipulation or agreement shall be made with any State which has not provided by law for a system of forest-fire protection: Provided further, That in no case shall the amount expended in any State exceed in any fiscal year the amount appropriated by that State for the same purpose during the same fiscal year. SEC. 3. That there is hereby appropriated, for the fiscal year ending June thirtieth, nineteen hundred and ten, the sum of one million dollars, and for each fiscal year thereafter a sum not to exceed two million dollars for use in the examination, survey, and of lands located on the headwaters of navigable streams or those which are being or which may be developed for navigable purposes: Provided, That the provisions of this section shall expire by limitation on the thirtieth day of June, nineteen hundred and fifteen (Sections 4 and 5—Repealed by P.L. 94–588).

Source: Sixty-first Congress. Sess. III. Chs. 181, 185, 186. (1911). https://www.govinfo.gov/content/pkg/STATUTE-36/pdf/STATUTE-36-Pg961-2.pdf/.

***Fire Suppression*, Roy Headley (1916)**

Not well known except in the annals of US Forest Service documents, Roy Headley's research and publications had a profound influence on the development of the agency's policies and procedures. He entered the USFS in 1907 as a Deputy Ranger on the Coeur d'Alene Forest, quickly advancing to become Assistant Supervisor to organize the Palouse Division, serving as Supervisor of the Cabinet Forest. In late 1908, he accepted a job as Assistant Chief of Operations in the California District Office when decentralization began to take place, writing "Not until 1910 did we even get a glimmer of understanding of what the fire problem meant." He was referring to the blazes later called "The Big Blowup" or "The Big Burn" that scorched 3 million acres in Monday, Idaho, and Washington over a two-day period. The fires led to a series of agency policies, federal statutes, and administrators who firmly supported the concept of total fire suppression. As head of the Division of Fire Control, Headley created one of the first fire suppression manuals, clearly identifying the role of firefighters and how fires were to be suppressed in District 5. In the following excerpt, which deals with Suppression Financial Policy, he outlines jurisdictional roles and expenditures, noting that the goal should be on preventing damage rather than the number of acres burned.

Suppression Financial Policy. The following principle from a recent work applies perfectly to fire suppression: "Perfect efficiency requires that the means employed for the accomplishment of a result shall be neither more nor less in quality, quantity or cost than is just sufficient for the purpose. The use of excessive means is only less unworkmanly than the use of insufficient means." The purpose of fire suppression is to prevent destruction of values owned by the Government and cooperators when destruction can be prevented at a cost less than the values in danger of destruction. From the foregoing may be deduced the following rules and principles which should govern in District 5:

1. The regular prevention organization should be used to suppress all fires occurring inside the Forest boundary, except on lands owned by noncooperating owners of private lands the owners of which are prepared to handle their own fires. When a reported fire is not to be fought by the Forest Service the owner of the land should be notified. Addresses of such owners should be kept on file by the Supervisor to permit prompt notification in the event of a fire.
2. Expenditures from the fire-fighting fund will be made with the sole object of protecting values owned by the Government and private owners giving satisfactory protective cooperation.

3. It is sometimes necessary to use the prevention forces and expand suppression funds on fires on private lands in order to suppress a fire which, if not controlled by the Forest Service, will spread until it must be suppressed at greater cost in order to prevent destruction of Government or cooperative values of consequence. Before expending suppression money on fires on lands owned by noncooperators or owners who should be able to handle their own fires, it should first be determined that by doing so the expense of protecting Government and cooperative values will be less than if some other plan is followed.
4. On the Santa Barbara, Angeles, Cleveland and Monterey Forests the values are public values even though a portion of a watershed is in private ownership. The ownership of land should therefore have no bearing on the organization of suppression jobs. Different suppression plans are required for different types of country, in that the foregoing policies may be applied. Each Forest should be studied to determine just what degree of intensiveness should be used under different conditions. For example, it should be pre-determined how much money, if any, will be spent in suppression fires on shotgun additions.
5. Effectiveness per man decreases rapidly as the size of a suppression crew increases, and usually cost per man increases with increase in numbers. The smaller the crew, therefore, the smaller the cost to construct and hold a given length of control line, but, of course, the longer the time.
6. The emphasis should be on prevention of damage rather than on prevention of burnt acres. Suppression jobs should be so organized always that the sum of damage to Government and cooperative values plus the cost of suppression will be at a minimum. A $500 suppression charge on a fire which destroyed $50 of values would be all wrong if a $300 suppression charge would have resulted in only $100 damage. There would be a difference of $150 in favor of the $300 plan of suppression. On the other hand, a $500 suppression plan on a fire with a $1000 damage would be all wrong if an $800 suppression plan would have resulted in a $600 damage. The difference would be $100 in favor of the $800 suppression plan. This principle frequently calls for the deliberate sacrifice of Government land in isolated tracts or on the fire side of a ridge or stream at which an inexpensive control line can be constructed or maintained.
7. Except on the Cleveland, Angeles, Santa Barbara and Monterey, no money should be spent to keep down burnt area in fields of pure brush. It is

necessary to be sure, however, that the stand is pure brush. Even very scattering reproduction means that the process of converting the brush cover into timber is started. Its destruction is frequently as much or more of a loss than the damage done by a fire in a mature stand.

8. Except in southern California fall and early spring fires burn lightly, flaring up only in spots. The damage is much less than by fires in the middle of the season. This calls for fewer men, less cost and a risk of larger acreage than would be acceptable in summer. The principle of minimum total damage and cost should be carefully applied. The seasonal factor must be considered.
9. Economics which mean larger burned acreage or delay in quenching fires should be modified more or less by the fact that in the event of a gale of wind the chance for a disastrous conflagration is increased by every acre bearing live fire. The time and character of the day must be considered.
10. Any men engaged on any fire should be worked to a reasonable limit. The danger of damage may not be sufficient to warrant the excessive effort expected on bad summer fires, but no fire is harmless enough to permit taking it easy. Dawdling on patrol instead of steady work cutting snags and covering fire may be responsible for much useless expense.

Source: Headley, Roy. 1916. *Fire Suppression, District 5.* United States Forest Service. https://firescope.caloes.ca.gov/SiteCollectionDocuments/Fire_Fighting%201916_style_ja_1916_headley.pdf.

The Uncontrollable Fire, Roy Headley (1919)

In the following document, Roy Headley makes the case for forest protection by "improvement" and argues that the loss of timber in California's national forests was progressively being reduced because of the agency's policies.

Some men say that the Government is all wrong in keeping fire out of the National Forests, that keeping fire out. Simply means piling up an accumulation of inflammable brush and litter, and that sooner or later a fire is bound to get started in this accumulation and sweep all before it.

There is a certain amount of truth in this theory. Fires which are allowed to run at will over a country undoubtedly consume large quantities of the dead wood and vegetable matter on the ground, and in addition much living brush and many small trees are killed or destroyed. Sometimes the fire leaves dead but unconsumed material which makes the area more inflammable than ever, and sometimes the fire kills even the mature timber and brings about the growth of an impenetrable brush field; but in general, it is true of a large acreage (particularly in yellow pine) that fire keeps the ground "clean" so one can ride around over it on horseback and see long distances.

On the other hand, when the Forest Service keeps fires out, the litter certainly accumulates; the brush, at least where former fires have left insufficient mature trees to shade it out, gets worse; and in addition to the litter and the brush, there comes up usually—a rank growth of tree seedlings—often occurring in dense thickets. All this growth seems like "brush" to the man who has to ride through it after cattle, but the forester distinguishes very carefully between brush and tree seedlings, for the same reason that the grain farmer distinguishes between weeds and wheat.

Now all this litter and brush and tree reproduction unquestionably makes it harder to stop a fire than if the ground had been kept "clean"? by repeated burnings. What, then, shall the Government do? If its main object was to keep the ground clear, and if there were large areas of agricultural lands to be cleared, the burning which the Indians and early Spaniards are said to have practiced might secure excellent results. Its object, since all large areas of agricultural lands are eliminated, is to make the National Forests produce as near as practicable to 100 percent of their maximum timber-producing' capacity, instead of carrying, as they now do, a mere fraction of the possible stand. We have now on the National Forests an average of, say, fifteen mature trees per acre, while we should and can have at least twenty-five. One reason we have fifteen instead of twenty-five is—fire. The old Spanish and Indian method of keeping the ground clean was, of course, a poor method of getting a full crop of timber. A look at one of the brush fields, so common in the timber belt in northern California, on most of which there is evidence that fire killed the original good stand of timber, will prove this to one's satisfaction. Whenever we see a brush field with dead snags still standing in it, we may know that the process of converting timberland into brush fields by fire is going on there right under our eyes.

If, then, the purpose of National Forest management is to increase the density of the stand of timber and to bring the annual income of salable timber to a maximum at the earliest practicable date, what is to be done about the increasing difficulty of protecting such a forest from fire? There are sincere people who believe that, because of the dense growth which has accumulated and will accumulate as a result of keeping fires out, a fire will sometime occur which no human power can control. Is it true that such a fire may sweep away not only the tree reproduction so much desired, but also a lot of the mature and merchantable timber which would have survived but for the practice of keeping fires out? And, if so, why try to improve our forests if the effort will merely lead eventually to the destruction of such forests as we now have? There are, no doubt, pessimists in San Jacinto and Hemet who say there is no use to rebuild anything but shacks, because, sooner or later, earthquakes will wreck any first-class buildings they may put |up. Nevertheless, the towns of San Jacinto and Hemet have already been rebuilt, and with first-class structures, for that is the spirit which has made California and which has made the Pacific Coast.

Would anyone seriously propose that the Forest Service abandon the effort to improve the forests, because the very improvement of the forests would make fire protection more and more difficult? "But," someone may say, "building a fully productive forest and building cities are not the same thing." "I am still afraid," this doubter may say, "that fire cannot be controlled as the litter and | other material accumulate." It happens that facts have been recently brought out that bear directly on this point. If the theory is true that fires cannot be controlled after protection has allowed litter, etc., to accumulate, then it is reasonable, is it not, to expect that losses from fire would increase as the years go—by and the inflammable material piles up? And if it should be found that the contrary actually happens and that, as the litter and reproduction increase with passing years of fire protection, the losses from fire actually decrease—then should not the prophecy of failure for fire protection stand discredited; and should not the burden of proof fall on the pessimist who questions the ability of the Forest Service to keep out fires permanently? Here is what has actually happened.

Systematic fire protection by the Forest Service on the National Forests in California began in 1905 and 1906, about twelve years ago. In this 12-year period an enormous amount of litter has accumulated; brush has grown to full density; and tree reproduction has thrived in many parts of the country until the very face of nature is altered. The greater part of the fire losses on the National Forests during this period have occurred in the two bad fire years—1910 and 1917.

In comparison with 1910, the season of 1917 was not only the longest and dryest; it was a harder fire season because seven more years of litter, brush, and seedlings had accumulated. Everyone knows how much stuff grows in seven years. The year 1917 should therefore have witnessed much more severe losses if there is any truth in the pessimist's theory that fire protection cannot protect permanently. What actually happened was this: In 1910 the area of timberland burned over in the California National Forests was 240,000 acres. In 1917 the area of timberland burned over was 20 percent less. In 1910 the merchantable timber destroyed amounted to 116 million feet. In 1917 only 22 million feet were destroyed—a decrease of 81 percent.

In 1918 there was another bad fire season, closely comparable in many respects with the 1910 and 1917 seasons; but the area burned over was only 73,000 acres and the timber destroyed was only 20,000,000 feet. In short, losses are being progressively reduced as the years go by.

In the face of this record is there any ground for saying that protection cannot protect permanently? The Forest Service asks the earnest support of every loyal citizen in keeping down fire losses. Every individual can give material help by getting the safety habit with fire and by urging others to do the same. Forest and county officials will appreciate it if you will offer to aid them in enforcing the State and Federal fire laws.

Source: Headley, Roy. 1919. *The Uncontrollable Fire.* California: United States Department of Agriculture, Forest Service. https://archive.org/details/uncontrollablefi00head/page/n1/mode/2up.

"Wildlife Management in the National Parks," The Leopold Report (1963)

In the early 1960s, rangers in Yellowstone National Park initiated two important policy changes. First, after years of feeding the park's extensive herds of elk, rangers decided to cull the herds that had eaten their way through the 2 million acres of vegetation to a point that was no longer considered sustainable. Predator control had eliminated their natural enemies, and megafauna in Yellowstone had flourished. During the winter of 1961–2, 4,283 elk were killed, causing a public outcry; the Department of the Interior responded by forming the Advisory Board on Wildlife Management. University of California professor A. Starker Leopold was selected as its chair, and the forthcoming document became known as The Leopold Report. While its ostensible mission was to discuss how best to deal with the elk problem, the document written by wildlife biologists changed course to examine the broader mission of national parks and how to manage them. The report, issued on March 3, 1963, and excerpted here, called for a return to national parks to their natural state, with a focus on the reintroduction of fire so that the landscape would resemble what it had looked like when first seen by Europeans. Fire suppression, the National Park Service's policy for decades, was viewed as the problem that had led to parks that were deteriorating, with naturally caused fires the solution, accompanied when appropriate by prescribed burning. The goal was ecological integrity built on a foundation of scientific research that included fire as the ultimate management tool.

Many of our national parks—in fact most of them—went through periods of indiscriminate logging, burning, livestock grazing, hunting, and predator control. Then they entered the park system and shifted abruptly to a regime of equally unnatural protection from lightning fires, from insect outbreaks, absence of natural controls of ungulates, and in some areas elimination of normal fluctuations in water levels. Exotic vertebrates, insects, plants, and plant diseases have inadvertently been introduced. And of course lastly there is the factor of human use–of roads and trampling and camp grounds and pack stock. The resultant biotic associations in many of our parks are artifacts, pure and simple. They represent a complex ecologic history but they do not necessarily represent primitive America.

When the forty-niners poured over the Sierra Nevada into California, those that kept diaries spoke almost to a man of the wide-spaced columns of mature trees that grew on the lower western slope in gigantic magnificence. The ground was a grass parkland, in springtime carpeted with wildflowers. Deer and bears were abundant. Today much of the west slope is a dog-hair thicket of young pines, white fir, incense cedar, and mature brush—a direct function of overprotection from natural ground fires. Within the four national parks—Lassen, Yosemite, Sequoia, and Kings Canyon—the thickets are even more impenetrable than elsewhere. Not only is this accumulation of fuel dangerous to the giant sequoias and other mature trees but the animal life is meager, wildflowers are sparse, and to some at least the vegetative tangle is depressing, not uplifting. Is it possible that the primitive open forest could be restored, at least on a local scale? And if so, how? We cannot offer an answer. But we are posing a question to which there should be an answer of immense concern to the National Park Service.

Of the various methods of manipulating vegetation, the controlled use of fire is the most "natural" and much the cheapest and easiest to apply. Unfortunately, however, forest and chaparral areas that have been completely protected from fire for long periods may require careful advance treatment before even the first experimental blaze is set. Trees and mature brush may have to be cut, piled, and burned before a creeping ground fire can be risked. Once fuel is reduced, periodic burning can be conducted safely and at low expense. On the other hand, some situations may call for a hot burn. On Isle Royale, moose range is created by periodic holocausts that open the forest canopy. Maintenance of the moose population is surely one goal of management on Isle Royale.

In essence, we are calling for a set of ecologic skills unknown in this country today. Americans have shown a great capacity for degrading and fragmenting native biotas. So far we have not exercised much imagination or ingenuity in rebuilding damaged biotas. It will not be done by passive protection alone.

Source: Leopold, A. Starker et al. 1963. "Wildlife Management in the National Parks." Advisory Board on Wildlife Management. March 4. https://www.nps.gov/parkhistory/online_books/leopold/leopold.htm.

Fergus County, Montana, Community Wildfire Protection Plan (2024)

One of the tools communities now use to prepare and respond to wildfires resulted from the encouragement outlined in the Healthy Forests Restoration Act of 2003. Community Wildfire Protection Plans (CWPP) have been developed using a variety of processes and participants. As a result, no two are alike in format, content, or length. On the one hand, the vagueness of the legislation has allowed for creativity and flexibility in what communities include in their plans, from data about current conditions that could lead to a wildfire; to examples of challenges that hinder action; to inventories and timelines of mitigation measures; to various maps, surveys, and lists. Although case studies of CWPP illustrate the diversity of how they are prepared, who participates, and how they are then used, there are also commonalities, including the identification of fuels reduction priorities, recommendations for action, and the use of collaborative decision-making by stakeholders. Typically, there is a high degree of local engagement in the identification of hazard mitigation strategies, but surprisingly, plan participation and diversity are unrelated to wildfire hazard. These findings suggest that CWPPs have been largely successful in their intent to engage diverse stakeholders in preparing for and mitigating wildfire risk. This excerpt of a recent CWPP is from Fergus County in the geographic center of Montana, with a population of less than 12,000 loosely scattered over an area of 4,253 square miles. The county prepared its first CWPP in 2004; then updated it two decades later in a 73-page plan. The list of participants, which includes multiple volunteer fire departments, illustrates how a small rural community can develop a comprehensive plan that could be a model for much larger jurisdictions to examine.

In 2004, in collaboration with the Fergus County Wildland-Urban Interface Wildfire Mitigation Plan Committee which was initiated by the Fergus County Department of Emergency Services developed a Community Wildfire Protection Plan (CWPP) and incorporated input from numerous stakeholders to help residents, neighborhoods, and communities mitigate potential threats from wildfire with the goal of increasing wildfire resiliency and survivability. In the past twenty years, changes have occurred across the county, including new infrastructure like roads, powerlines, buildings, and housing. Additionally changes in fuels/vegetation have occurred including areas of conifer encroachment, land use changes that remove grazing, wildfires on the landscape, and forestry/fuels treatments near communities. These changes affect the way a community plans for fire and prompts the need for revisions. The purpose of a CWPP is to outline the risks and hazards associated with a wildland fire to Fergus County communities and to identify potential mitigation measures. The Fergus County Community Wildfire Protection Plan (CWPP) is intended to provide documentation of implementing actions designed to reduce risk to homes and communities from wildfire through education and outreach programs, the development of partnerships, and implementation of preventative activities such as hazardous fuel reduction, ignition source mitigation, defensible space, land use, and/or building codes. The emphasis of this plan is to work from the home outward into the wildland urban interface, so that man-made and natural resources survive the intrusion of a wildfire. This plan is intended to:

1. Meet the requirements of the Healthy Forest Restoration Act (HFRA) of 2003,
2. Make the County eligible for National Fire Plan (NFP) funding assistance from the Departments of Agriculture and Interior (by meeting the requirements of HFRA),
3. Provide information to assist communities in developing fuel reduction projects on private and public lands,
4. Continue to serve as the Wildfire Hazard Mitigation portion of Fergus County's Multi-Hazard Mitigation Plan. This is required for counties to be eligible to receive FEMA disaster assistance funding.

This 2024 update includes efforts to create a more streamlined plan and a complete review of the mitigation recommendations to remove projects that may no longer be relevant, and to modify or add projects where changes have occurred. This plan update includes new information, renewed emphasis on the role of embers in wildfire spread, and potentials of climate change. This plan updates information on historic large fire starts in Fergus County, weather data, and fire ignition sources. This plan also includes updates to the fire district boundaries and staffing. The Montana Department of Natural Resources and Conservation (MT-DNRC) and USDI Bureau of Land Management (BLM) provided funding for development of this plan. The successful development of the 2024 CWPP was made possible only with the active support and assistance of many people who devoted countless hours to the project. These included citizens, MT-DNRC, BLM, Fergus County employees and supervisors, local fire chiefs, and Snowy Mountain Development employees.

Key Stakeholders and Collaboration

The Fergus County Commissioners, working collaboratively with the Montana Department of Natural Resources and Conservation and USDI Bureau of Land Management, solicited competitive bids from companies to provide the services of leading the plan update and writing the Fergus County Community Wildfire Protection Plan (CWPP). The commissioners initially selected HydroSolutions to lead the effort and two stakeholder meetings were held to discuss roles and responsibilities. During the stakeholder meetings, Fergus County identified Snowy Mountain Development to lead the planning effort and transitioned the technical writing of the plan to them.

Key stakeholders included:

Infrastructure

- NorthWestern Energy (Powerlines and Natural Gas Pipeline)
- Fergus Electric Cooperative (Powerlines)
- Centurylink Communications
- Triangle Communications
- MidRivers Communications
- T-Mobile (Cell Phone Towers)
- Verizon Wireless (Cell Phone Towers)
- AT&T Wireless (Cell Phone Towers)

- City of Lewistown (Municipal Water Supply for Hydrant System within City Limits) Department of Defense (DOD) Missile sites and command centers
- Oil tanks
- Railroad

Schools

- Fergus County School Superintendent
- Lewistown Public Schools
- Grass Range Public Schools
- Moore Public School
- Roy Public School
- Winifred Public School
- King Colony Public School
- Denton Public School

Emergency/Fire Response

- Fergus County Fire Council
- Beaver Creek/Cottonwood Volunteer Fire Department
- Cheadle Volunteer Fire Department
- Coffee Creek Volunteer Fire Department
- Denton Volunteer Fire Department
- Grass Range Volunteer Fire Department
- Heath Volunteer Fire Department
- Hilger Volunteer Fire Department
- Lewistown Fire
- Lewistown Rural Fire
- Moore Volunteer Fire Department
- North Fork Flatwillow Volunteer Fire Department
- Roy Volunteer Fire Department
- Winifred Volunteer Fire Department
- Fergus County Sheriff's Office

Federal Land Management

- USDA Forest Service—Lewis and Clark National Forest, Judith Ranger District

- USDI Bureau of Land Management, North Central Montana District, Lewistown Field Office, and Upper Missouri River Breaks National Monument
- USDI Fish and Wildlife Service, Charles M Russell National Wildlife Refuge

State Land Management

- Montana Department of Natural Resources and Conservation
- Montana Fish, Wildlife and Parks

Other Land Management Organizations

- Fergus County Conservation District
- Big Spring Watershed Group
- American Prairie Reserve

Other Stakeholders

- Lewistown Insurance
- Montana Farm Bureau Insurance
- State Farm Insurance

Connection to Plans and Agreements in Place

Fergus County has cooperative agreements for mutual aid fire response with the following counties: Pondera, Chouteau, Teton, Toole, Liberty, Glacier, Judith Basin, Wheatland, Golden Valley, Petroleum, Phillips, Blaine, Cascade, and Hill. The County also has agreements with the Blackfeet, Rocky Boy, and Fort Belknap Nations. Fergus County has firefighting mutual aid agreements with the Montana Department of Natural Resources and Conservation and has a general mutual aid agreement with the USDI Bureau of Land Management. All fire districts within Fergus County are represented at the Fergus County Fire Council and mutual aid automatic agreements have been signed by the districts.

Source: Fergus County, Montana, Community Wildfire Protection Plan. 2024. https://dnrc.mt.gov/forestry/Community-Local-Government/2024-Fergus-County-CWPP-Signed.pdf.

Executive Order N-4-25, California Governor Gavin Newsom (2025)

The series of wildfires that devastated parts of Los Angeles County in January 2025 were accompanied by numerous pronouncements and statements ranging from ones issued by the President of the United States to others from mayors, utility spokespersons, and program and department administrators, among others. The documents provide a written chronology as well as narratives in real time that offer differing perspectives about the causes and effects of the blazes. Just days after the first fire began, California Governor Gavin Newsom signed an executive order announcing the suspension of the permitting and review requirements under California's Environmental Quality Act (CEQA) and the California Coastal Act. The purpose of the order, issued January 12, 2025, was to streamline the rebuilding of homes and businesses destroyed in the firestorms. The executive order also extended protections against price gouging for building materials, storage services, construction, and other essential services for one year. The governor pledged to work with the state legislature to identify statutory changes that could help expedite rebuilding while enhancing wildfire resilience and safety—broad language with little specificity. The governor's N-4-25 executive order was issued in addition to one where he proclaimed a State of Emergency in two counties (January 7), two others to bolster the emergency response to the spread of the fires (January 8 and 10), and another (January 27) directing the California Coastal Commission not to take actions that would conflict with the governor's other executive orders. In addition, the January 27 document allowed fire survivors to stay in hotels or other short-term rentals for more than thirty days to boost the area's short-term housing availability. The flurry of announcements streamlined existing policies but also kept the governor's profile front and center at a time when other political figures were also clamoring for attention. Executive orders like these tend to be broad outlines of policy preferences, rather than specific directives. While short on detail, they allow an official to identify key issues and claim credit for doing so, leaving the heavy lifting of policy implementation to managers and administrators. In some cases, the executive order will later be canceled or rebuffed after judicial review or in response to public opposition. More often, administrative discretion will be used in trying to translate what the official meant into equitable, affordable, or acceptable implementation.

WHEREAS on January 7, 2025, I proclaimed a State of Emergency to exist in Los Angeles and Ventura Counties due to fire and windstorm conditions that have now caused multiple fires, including the Palisades, Eaton, Hurst, Lidia, Sunset, and Woodley Fires; and WHEREAS on January 8, 2025, and January 10, 2025, I issued Executive Orders to bolster the emergency response to the rapid spread of these fires throughout Los Angeles and Ventura Counties; and WHEREAS these fires have collectively burned over 39,000 acres, and these fires and windstorm conditions have devastated communities across the Greater Los Angeles Area and destroyed or damaged more than 12,000 structures, including homes and businesses, with initial estimates placing this emergency among the most destructive in California history; and WHEREAS Los Angeles County is home to over 9.6 million people and constitutes one of the most vibrant and productive regions in the United States; and WHEREAS the Los Angeles region has some of the highest housing costs in the country, with recent studies finding that a majority of Los Angeles County renters are burdened by paying more than 30 percent of their income on rent; and WHEREAS the destruction of homes by the fires burning in Los Angeles and Ventura Counties will reduce the available housing in the region, which is likely to further increase the cost of housing and the related crisis of homelessness if housing is not expeditiously rebuilt; and WHEREAS I am committed to helping families and individuals who have lost their homes and businesses rebuild and recover without facing unnecessary costs and delay; and WHEREAS the State can expedite recovery from this disaster by eliminating barriers that unduly delay the rapid rebuilding of homes and other facilities destroyed or damaged by the extreme windstorm conditions and resulting fires; and WHEREAS efforts to rebuild should include measures to increase community resilience, harden homes, and ensure defensible space to build resilience to future wildfires, to the greatest extent practicable; and WHEREAS current law provides for certain waivers of state permitting and building codes, but additional streamlining is necessary to facilitate a rapid and resilient rebuild; and WHEREAS under Proposition 13, and the Revenue and Taxation Code, the base year value of property that is substantially damaged or destroyed by a disaster, as proclaimed by the Governor, may be transferred to rebuilt property; and WHEREAS my January 7, 2025, Proclamation of a State of Emergency triggered a California law that automatically restricts price gouging for building materials, housing and storage services, emergency supplies, and repair, reconstruction, and emergency clean-up services when a state of emergency is proclaimed, but those protections expire after 30 to 180 days, depending on the good or service, and rebuilding efforts in the areas devastated by these fires will extend beyond that time; and WHEREAS under the provisions of Government Code sections 8571, I find that strict compliance with various statutes and regulations specified in this Order would prevent, hinder, or delay the mitigation of the effects of these fires and windstorm conditions. NOW, THEREFORE, I, GAVIN NEWSOM, Governor of the State of California, in accordance with the authority vested in me by the State Constitution and statutes, including the California Emergency Services Act, and in particular, Government Code sections 8567, 8571, and 8627, do hereby issue the following Order to become effective immediately. IT IS HEREBY ORDERED THAT: 1. With respect to projects to repair, restore, demolish, or replace property or facilities substantially damaged or destroyed as a result of this emergency, and that satisfy the requirements provided in Paragraph 2, the following are suspended to the extent they would otherwise apply: a. The California Environmental Quality Act (Division 13 of the Public Resources Code, commencing with section 21000), and regulations adopted pursuant to that division; and b. Any requirement to obtain a permit under the California Coastal Act (Division 20 of the Public Resources Code, commencing with section 30000). 2. Paragraph 1 shall apply only to properties and facilities that are in substantially the same location as, and do not exceed 110% of the footprint and height of, properties and facilities that were legally established and existed immediately before this emergency. 3. The provisions of Penal Code section 396, subdivisions (b) and (c), prohibiting price gouging in times of emergency, shall remain in effect until January 7, 2026, for Los Angeles County. The time limitations under those subdivisions are hereby suspended. 4. The Department of Housing and Community Development (HCD), the Office of Land Use and Climate Innovation, the Office of Emergency Services, and the Department of General Services (DGS) shall, within 30 days, provide a report to me identifying other state permitting requirements that may unduly impede efforts to rebuild properties or facilities destroyed as a result of this emergency that should be considered for suspension, and shall update that report every 60 days, as appropriate and as recovery and rebuilding efforts proceed, to identify any additional permitting requirements that are posing barriers to rebuilding and that should be considered for suspension. 5. HCD, in consultation with DGS, the Office of the State Fire Marshal, and the California Energy Commission, shall, within 60 days, review and provide a report to me with recommendations regarding any provision of the Building Standards Code, Title 24 of the California Code of Regulations, that should be suspended for projects described in Paragraphs 1 and 2 in order to facilitate rapid, safe, and cost-effective rebuilding and recovery. 6. HCD shall coordinate with local governments to identify and recommend procedures, including but not limited to exploring the use of preapproved plans and waivers of certain permitting requirements, to establish rapid permitting and approval processes to expedite the reconstruction or replacement of residential properties destroyed or damaged by fire. The recommended procedures shall have the ultimate goal of issuing all necessary permits and approvals within 30 days. HCD shall, within 60 days, provide a report to me identifying recommended updates to local government procedures that achieve these goals, and shall update that report every 60 days, as appropriate, to identify any additional permitting and approval requirements that are barriers to recovery and rebuilding efforts that should be removed. 7. My Administration shall collaborate with the Legislature to identify and propose statutory amendments that durably address barriers impeding rapid rebuilding efforts in the areas affected by this emergency, including potential legislation to incentivize the incorporation of additional fire hardening measures into rebuilding efforts and enhancements of fire mitigation and fire response capacity within rebuilt areas, and potential legislation based on any suspensions of statutes and regulations in this Order, or any subsequent order, or reports required under Paragraphs 4 through 6. I FURTHER DIRECT that as soon as hereafter possible, this Order be filed in the Office of the Secretary of State and that widespread publicity and notice be given of this Order. This Order is not intended to, and does not, create any rights or benefits, substantive or procedural, enforceable at law or in equity, against the State of California, its agencies, departments, entities, officers, employees, or any other person. IN WITNESS WHEREOF I have hereunto set my hand and caused the Great Seal of the State of California to be affixed this 12th day of January 2025.

Source: Executive Department. State of California. Executive Order N-4-25. 2025. https://www.gov.ca.gov/wp-content/uploads/2025/01/EO-N-4-25-Rebuilding-Final-signed.pdf.

Hawai'i Supreme Court Order: Maui Fire Cases (2025)

The short order announced by the Supreme Court of Hawai'i on February 10, 2025, is in sharp contrast to the August 8, 2023, wildfire in Lahaina, Maui, that became the nation's deadliest, causing an estimated $5.5 billion in damage. In the eighty-page formal decision issued a month later, the court's opinion closed the loop in the $4.04 billion settlement litigation in actions filed by wildfire victims against the seven defendants accused of causing the tragedy. The main defendant, Hawaiian Electric, and the state of Hawai'i, Maui County, and Kamehameha Schools had been sued because of their role in starting and allowing the fire to spread, and had agreed to pay the settlement in a deal reached in 2024. More than a thousand people had filed hundreds of suits in the case.

The focus of the case was the question whether Hawai'i law allowed the 192 insurers to go after the parties that agreed to fund the settlement (the defendants), and to instead seek liens against the insurance policyholders—the victims who filed claims and were eligible for a portion of the $4 billion settlement that had been reached earlier. The insurers, who had already paid out $2.3 billion in claims to the victims, argued that they could recoup their losses as the "insurer's exclusive remedy." That remedy is called the right of subrogation, and it allows insurance companies to recover the money they had paid to their policyholders. In ruling against the insurers, the court sent the case back to the Maui Circuit Court to oversee the administrative process of deciding how much each victim will receive. Lawyer fees could potentially amount to more than $1 billion. Still unknown is whether the insurers will raise their insurance rates or pursue liens against the customers who held policies, or leave the state entirely, as some have done elsewhere after massive wildfires .

SCRQ-24-0000602 IN THE SUPREME COURT OF THE STATE OF HAWAI'I IN THE MATTER OF THE PETITION FOR THE COORDINATION OF MAUI FIRE CASES. RESERVED QUESTIONS FROM THE CIRCUIT COURT OF THE SECOND CIRCUIT (CASE NO. 2CSP-23-0000057) ORDER (By: Recktenwald, C.J., McKenna, Eddins, and Ginoza, J.J., and Circuit Judge Morikone, in place of Devens, J., recused).

The Circuit Court of the Second Circuit reserved three questions to this court. Upon consideration of the appellate briefing, the record, and oral argument, IT IS HEREBY ORDERED that the reserved questions are answered as follows: Question 1: Does the holding of Yukumoto v. Tawarahara, 140 Hawai'i 285, 400 P.3d 486 (2017), that limited the subrogation remedies available to health insurers to reimbursement from their insureds under HRS § 663-10 and barred independent actions against tortfeasors who settled with the insureds extend to property and casualty insurance carriers? Electronically Filed Supreme Court SCRQ-24-0000602 10-FEB-2025 09:09 AM Dkt. 192 ORD 2 Question 1 is answered in the affirmative. Our opinion in Yukumoto v. Tawarahara, 140 Hawai'i 285, 400 P.3d 486 (2017), extends to property and casualty insurers such that, under Hawai'i Revised Statutes (HRS) § 431:13-103(a)(10)(A), the lien provided for under HRS § 663-10(a) is the exclusive remedy for a property and casualty insurer to recover claims paid for damages caused by a third-party tortfeasor in the context of a tort settlement between an insured and the tortfeasor. Question 2: Is a property and casualty insurer's subrogation right of reimbursement prejudiced by its insured's release of any tortfeasor when the settlement documents and release preserve those same rights under HRS § 663-10? Because the statutory lien under HRS § 663-10 is the exclusive remedy for a property and casualty insurer in the context of a tort settlement, Question 2 is answered in the negative. Question 3: Under the circumstances of the Maui Fire Cases and the terms of the "Global Settlement," does the law of the State of Hawai'i require that insureds be made whole for all claimed injuries or damages before their insurers can pursue a subrogation right of recovery or reimbursement against a third-party tortfeasor? Question 3 is answered in the negative. Under the circumstances of this mass tort case, we decline to apply the made whole doctrine to the statutory lien-claim process established by HRS §§ 431:13-103(a)(10) and 663-10. 3 IT IS FURTHER ORDERED that this case is remanded to the circuit court for such other and further proceedings as may be appropriate consistent with this order. We retain concurrent jurisdiction to enter an opinion and judgment that will follow. DATED: Honolulu, Hawai'i, February 10, 2025. /s/ Mark E. Recktenwald /s/ Sabrina S. McKenna /s/ Todd W. Eddins /s/ Lisa M. Ginoza /s/ Kevin T. Morikone.

Source: State of Hawai'i Supreme Court. 2025. https://www.courts.state.hi.us/wp-content/uploads/2025/02/SCRQ-24-0000602ord.pdf.

6

Resources

Books

Alcock, John. 2017. *After the Wildfire: Ten Years of Recovery from the Willow Fire.* Tucson: University of Arizona Press.

The author started documenting the damage done by the 2004 Willow Fire less than a year after the fire started. A naturalist, he examines ten years of what changed in the landscape of central Arizona's Mazatzal Mountains and Deer Creek, noting which species survived, those that did not, and the growth in plant life that then promotes the biodiversity of the region. There is more detail than the average reader might like, with a writing style that is more poetic than scientific.

Anderson, M. Kat. 2013. *Tending the Wild: Native American Knowledge and the Management of California's Natural Resources.* Berkeley: University of California Press.

Few of those who have written about traditional native American practices related to wildfire have relied upon the oral histories and interviews conducted by the researchers themselves. This book is an exception. Anderson presents a comprehensive picture of California under the management of its original residents.

Arno, Stephen F., and Steven Allison-Bunnell. 2002. *Flames in Our Forests: Disaster or Renewal.* Washington, DC: Island Press.

Damage from the 1988 wildfire, caused by multiple lightning strikes, that burned 683,000 acres in Yellowstone National Park, Wyoming. The fire brought into question the Department of Interior's policy that allowed naturally caused wildfires to burn if they do not threaten human life, property, or historic sites (Tamifreed/Dreamstime.com).

Although this book is nearly twenty-five years old, it provides an excellent historical perspective on issues that still have not been resolved, such as controlled (now prescribed) fire, fire suppression, and the role of politics and bureaucracy. The authors note that the credibility of the US Forest Service is low (and still is now), with many of the agency's initiatives thwarted by litigation from outside groups. This is an excellent source for historical comparisons.

Barker, Rocky. 2007. *Scorched Earth: How the Fires of Yellowstone Changed America.* Washington, DC: Island Press.

In 1988, Barker was an environmental journalist covering the fires that raged in Yellowstone National Park, and in his book he deals not only with what he saw on the ground, but also with the historical context of how the country chose to preserve its public lands after the Civil War.

Belden, Anne E., and Paul Gullixson. 2023. *Inflamed: Abandonment, Heroism, and Outrage in Wine Country's Deadliest Firestorm.* New York: Permuted Press.

This is a detailed narrative of the Tubbs Fire on October 9, 2017, and the response by public safety officials and others living in Santa Rosa, California. The book's focus is on two senior living communities whose residents' lives were at stake, those who rescued them from the fire, and those who abandoned them. The story is riveting and would make a great screenplay.

Blackburn, Thomas C., and M. Kat Anderson, eds. 1993. *Before the Wilderness: Environmental Management by Native Californians.* Menlo Park, CA: Ballena Press.

This edited volume contains work by some of the most well known researchers who have studied native peoples' use of fire. Of particular interest is the selection by Henry T. Lewis, considered to be one of those who broke this topic open with his publications in the early 1980s.

Boyd, Robert, ed. 2022. *Indians, Fire, and the Land of the Pacific Northwest.* Corvallis: Oregon State University Press.

This updated version of the original book is a series of essays by noted researchers who write that the land encountered by the early explorers was not the pristine Eden they expected to find. Rather, the land was carefully managed by the native people who used fire primarily to find food for their tribes. Traditional ecological knowledge (TEK) is referred to by several of the authors, providing a rare historical perspective to Indian practices.

Bradshaw, Karen M., and Dean Lueck, eds. 2012. *Wildfire Policy: Law and Economics Perspectives.* New York: Resources for the Future Press.

There are few books available that cover the topics in this edited volume, which range from the Australian experience with wildfire to common law liability and fire suppression. Although it is somewhat dated, the authors do raise issues and ask questions that are seldom analyzed elsewhere within either the scholarly or popular press. The hardcover version of this book is expensive, primarily because it was written as a scholarly approach to the topics in its title.

Butler, Joyce. 2014. *Wildfire Loose: The Week Maine Burned.* Rockport, ME: Down East Books.

In this reissue of the original 1979 edition, Butler tells a tale that others note reads like fiction but is, unfortunately, deadly true. The October 1947 wildfires impacted thirty-five communities in Maine, and she writes not only about the property damage, but more specifically, about the farmers whose livelihoods were lost, deer that suffocated in the blaze, and the fifteen people who perished. This is one of several books about Maine she has authored, but it goes beyond a yarn about Maine to identify the real losses that could apply to any community devastated by wildfire.

Cronon, William. 2003. *Changes in the Land: Indians, Colonists, and the Ecology of New England* (Revised edition). New York: Hill and Wang.

Historian Cronin's original version of this book is similar to the revised edition. Both editions paint a dark picture of the impact of British colonialism in America that is different from the traditional image of Pilgrims and savages. The context is New England, but the implications cross geographic boundaries.

Gates, Jon Humboldt. 2019. *Firestorm: A Personal Narrative from the Epicenter of a California Wildfire.* Lake Oswego, OR: Moonstone Publishing.

A native of northern California, where the Tubbs Fire took place in 2017, Gates and his family were awakened near midnight when they fled their home in one of the state's most destructive fires. This account is more personal and emotionally driven than others that chronicle the same fire with a journalistic approach.

Gess, Denise, and William Lutz. 2002. *Firestorm at Peshtigo.* New York: Henry Holt.

Often referred to as the country's deadliest fire, it is estimated that more than 2,200 people were killed in the October 8, 1871, blaze that raged over 2,400 square miles of forest in eastern Wisconsin. Reading almost like historical fiction, the authors focus on the human dimension of the fire, using newspaper accounts, archival materials, personal diaries, and letters to tell the story of how the city that provided the lumber to build Chicago burned—ironically—on the same day as the more highly publicized Great Chicago Fire.

Hudson, Mark. 2011. *Fire Management in the American West: Forest Politics and the Rise of Megafires.* Boulder: University Press of Colorado.

Written from what some might call a neo-Marxist perspective, the author, a sociologist, argues that the US Forest Service's policies of wildfire suppression are less to blame for the rise of larger, more intense fires than capitalism and the timber industry.

Jensen, Sara E., and Guy R. McPherson. 2008. *Living with Fire: Fire Ecology and Policy for the Twenty-First Century.* Berkeley: University of California Press.

The authors argue that wildland fire is a problem that should be mitigated, rather than solved. They suggest that media coverage that began with the Yellowstone fires of 1988 made wildfires a national media event, with newly developed tools like satellite news trucks and communication satellites making on-the-spot coverage affordable for smaller local stations. This allows the media to use images of fires to reduce wildfires to sound-bite status.

Linden, Eugene. 2022. *Fire and Flood: A People's History of Climate Change from 1979 to the Present.* New York: Penguin Press.

This book is more about climate change than about wildfire, but it does explain why fire and floods, and less directly, the insurance industry, are key to understanding our response to the problem. He puts the blame on a number of stakeholders, from the federal government to fossil fuel advocates.

MacEachern, Alan. 2020. *The Miramichi Fire: A History.* Montreal: Mc-Gill-Queen's University Press.

The author is an environmental historian who brings depth to the story of the not-well-known fires that burned in Maine and the Canadian colony of New

Brunswick on the same day in 1825. He uses primary sources from the time to explain the confusion that continues to blur facts and misinformation about the fires' causes and coverage of what the press said happened at the time.

Maclean, Norman. 1992. *Young Men and Fire.* Chicago, IL: University of Chicago Press.

Completed by others after his death, the book is reportorial in style, but it reads like the prose of a storyteller. Maclean tells of his obsession with Montana's August 1949 Mann Gulch Fire and the tragic death of thirteen US Forest Service smokejumpers. The book is considered a classic in wildfire history.

Maclean, John N. 2003. *Fire and Ashes: On the Front Lines of American Wildfire.* New York: Henry Holt.

Maclean's critics accuse him of sensationalizing wildfire, describing the scenes of firefighters' deaths in too much detail in order to sell books, of which this volume is one of many. Those who follow the author like his attention to detail and how he researches his subjects carefully. The prose does not seem to change, though, from one book to another; only the name and location of the fire differs.

Maclean, John N. 2009. *Fire on the Mountain: The True Story of the South Canyon Fire.* Boston, MA: Mariner Books.

For those who like to read non-fiction and first-person accounts about disasters, this book will be just what you are looking for. A string of mistakes and human errors led to the death of fourteen people in a 1994 Colorado wildfire where even the location (and thus the name) was wrong.

Martin, Manjula. 2024. *The Last Fire Season: A Personal and Pyronatural History.* New York: Pantheon Books.

Written as a memoir of the author's move to Sonoma County, she seeks to explain the increased frequency of wildfires in California in 2020. She describes what it was like for her to decide whether or not to evacuate, and how the fire affected her life afterward. The book is more personal than historical in tone.

McDonough, Brendan. 2017. *Granite Mountain.* New York: Grand Central Publishing.

Potential heroin addict turned wildland firefighter, the author's claim to fame is that he was the only survivor of the 2013 Granite Mountain Fire that claimed the lives of nineteen men near Yarnell, Arizona. Now a public speaker, the book is sometimes hard to read as he shares the reasons for his survivor's guilt and PTSD.

Mott, Nick, and Justin Angle. 2023. *This Is Wildfire: How to Protect Yourself, Your Home, and Your Community in the Age of Heat.* New York: Bloomsbury.

This is a practical reference book, designed to provide guidance on how individuals can implement ways to protect themselves, their family, and their homes against the growing threat of wildfire. The authors note that the threat of climate change, and more people moving into fire-prone areas, makes protection more critical than ever.

Pyne, Stephen J. 2004. *Tending Fire.* Washington, DC: Island Press.

The author's analysis of fire is told from the concept of three strands: earthly fire (planetary burning); European expansion and fire-prone public lands; and the evolution of American fire-related institutions in comparison with other nations. Some of Pyne's observations may now seem prescient, while others have less value in light of current events.

Pyne, Stephen J. 2010. *America's Fires: Management of Wildlands and Forests.* Durham, NC: Forest History Society.

In this revised edition, Pyne wears a historian's hat, placing some of the key changes related to wildfire policy in context, with particular attention paid to exclusion and suppression, and how this led to even more damaging fires.

Pyne, Stephen. 2015. *Between Two Fires: A Fire History of Contemporary America.* Tucson: University of Arizona Press.

Unlike some of his other books, Pyne writes this volume as an examination of wildfire policy starting with the 1960s and the shift from fire suppression to reform in the 1990s, and a management mission that he refers to as "evolving."

Reinhart, Karen Wilding. 2008. *Yellowstone's Rebirth by Fire: Rising from the Ashes of the 1988 Wildfires.* Helena, MT: Farcountry Press.

This could be called a coffee table book because of its engaging photographs, but it is more than that. The author provides a detailed timeline of the 1988 fires, and the photographs dramatically illustrate how the landscape that many believed was gone forever was, in fact, still very much alive. The book's maps help explain the spread of the fires through the park in a way that is not seen elsewhere.

Scott, Andrew C. 2018. *Burning Planet: The Story of Fire through Time.* Oxford: Oxford University Press.

A geologist whose research focuses on coal as a way of learning about fire, Scott writes less about humans and more about fossils and charcoal in rocks from millions of years ago. The book is best used by those with a background or interest in geochemistry. The material does not rank close to Pyne's books.

Staff of the *Billings Gazette.* Rev. Ed. 1995. *Yellowstone on Fire!* Billings, MT: *Billings Gazette.*

The photographers and writers from a city 100 miles from the 1988 blaze tell the story that is in some ways more compelling than longer, more academic versions. The result is a compilation of the key issues of the Siege of 1988, from the controversial "let it burn" policy that was heavily criticized by many to the natural replanting process that has replenished the forest.

Stewart, Omer C., Henry T. Lewis, and M. Kat Anderson, eds. 2009. *Forgotten Fires: Native Americans and the Transient Wilderness.* Norman: University of Oklahoma Press.

This edited volume focuses on the work of Omer Call Stewart, a renowned American anthropologist whose research on indigenous peoples contradicted the noble savage stereotype of others. The textbook-style compendium of his papers explains how landscape-scale fire was used to alter wilderness areas throughout the United States, and the book is now considered a classic in the field.

Struzik, Edward. 2017. *Firestorm: How Wildfire Will Shape Our Future.* Washington, DC: Island Press.

While much of the coverage of the book deals with Canada, there is ample historical material dealing with fires in the United States. The storytelling is well documented, and the author does a good job suggesting that there is a new wildfire paradigm. The author's analysis, however, might be considered dated.

Vale, Thomas, ed. 2002. *Fire, Native Peoples, and the Natural Landscape.* (Second edition). Washington, DC: Island Press.

There are seven chapters in this edited book, each one written from a different regionalized perspective of the West. All seven relate to the question of the role of fire in changing the landscape of pre-European contact to analyze different fire regimes in the region.

Vaughn, Jacqueline, and Hanna J. Cortner. 2005. *George W. Bush's Healthy Forests: Reframing the Environmental Debate.* Boulder: University of Colorado Press.

One of the key developments in recent forest policy is the legislative and regulatory battle over what would eventually become the Healthy Forests Restoration Act of 2003. Framed as a strategy for preventing catastrophic forest fires like those that had consumed southern California at the time the debate over forest health was underway, the measure pitted timber interests against environmental ones, producing policies that changed the ways both forests and wildfires are managed.

Younger, Sandra Millers. 2023. *The Fire Outside My Window: A Survivor Tells the True Story of California's Epic Cedar Fire* (Second edition). Nyack, NY: SureShot Books Publishing.

The author is a survivor of the 2003 Cedar Fire in San Diego, a journalist, and public speaker, and she combines those talents in this narrative documentary of her experience. She also covers the event from the perspective of the first responders, capturing the view in a different voice and her own eyes.

Articles

Adler, Maxwell, and John Gittelsohn. 2025. "California Wildfires Threaten Home Values across State, S & P Says." *Insurance Business Magazine.* January 27. https://www.insurancebusinessmag.com/us/business-news/california-wildfires-threaten-home-values-across-state-sandp-says-522148.aspx.

Although the analysis is preliminary, the authors present intriguing information on the future of California real estate. The figures are likely to undergo significant changes once all the tallies from the January 2025 wildfires are available.

Anton, Mike. 2009. "At 65, Smokey Bear Is Still Fighting Fires." *Los Angeles Times*. July 24. https://www.latimes.com/science/la-me-smokeybear24-2009jul24-story.html.

Each year, the anniversary of the Smokey Bear campaign brings out articles such as this one that explain why the bear's legacy continues.

Araullo, Kenneth. 2025. "State Farm Pays Out Nearly $500 Million in California Wildfire Claims." *Insurance Business Magazine*. January 28. https://insurancebusinessmag.com/us/news/catastrophe/state-farm-pays-out-nearly-500-million-in-california-wildfire-claims-522307.aspx.

As one of the state's largest insurers, State Farm rushed to provide its policyholders with at least initial payouts resulting from the January 2025 fires. Although the claims process continues, this article provides a perspective on the industry impact.

Arno, Stephen F. 2023. "Fire-Adapted Western Forests: The Role of American Indians." *Fire Management Today* 81 (2): 8–12. fs.usda.gov/sites/default/files/fire-management-today/FMT-81-2.pdf/.

Historian Arno presents compelling evidence on the way in which early indigenous people were able to change the composition and look of western forests long before European-American explorers and settlers arrived.

Arno, Stephen F. 2023. "The 'Light Burning' Controversy." *Fire Management Today* 81 (2): 13–16. fs.usda.gov/sites/default/files/fire-management-today/FMT-81-2.pdf.

"Light burning," or what some called Paiute forestry, has several names in wildfire parlance, but it remains as controversial now as it was in the early days of fire management.

Bassler, Hunter. 2024. "'Watch Our State Burn': Trump's Threat to Withhold California's Wildfire Aid Angers State's Firefighting Force." *Wildfire Today*. September 15. https://wildfiretoday.com/2024/09/15/trumps-threats-to-block-california-wildfire-aid-if-elected-president-firefighter-reaction/.

Leaders among California firefighters believe that the state is being held hostage by the president as the administration seeks to move its own agenda forward instead of releasing funding for damage from the 2025 wildfires.

Baumhardt, Alex. 2024. "Oregon Homeowners Face Soaring Premiums, Few Property Insurance Options over Wildfires." *Oregon Capital Chronicle*. February 26.

https://oregoncapitalchronicle.com/2024/02/26/oregon-homeowners-face-soaring-premiums-few-property-insurance-options-over-wildfires/.

The 2020 Labor Day Fires throughout Oregon led to homeowner insurance policies or premiums that were either increased or canceled. The author states that Oregon insurance rates are now beginning to look like the situation in California and Florida.

Blakeley, Kiri. 2025. "California Flood Risk Exceeds Fire Concerns, according to Study." *Realtor.com*. February 19. https://www.realtor.com/news/trends/california-flood-risk-exceeds-fire-concern/study/.

Although it may not be much consolation to the victims of the January 2025 wildfires, this author notes that a more devastating hazard for Californians is flooding, especially after a fire.

Borenstein, Seth. 2025. "It's Not Really the Typical Time for Nasty California Fires. What Changed That?" *AP News*. January 9. https://apnews.com/article/fire-devastation-climate-change-santa-ana-winds/.

There is a debate over whether "fire seasons" exist, especially given what is becoming just as common—fires that occur through the year. This is particularly true, the article says, in the West.

Briscoe, Tony. 2025. "FEMA Doubles Down on Its Decision to Not Test Soil as Part of Wildfire Cleanup." *Los Angeles Times*. February 15. https://latimes.com/environment/story/2025-02-15/fema-defends-fire-cleanup-strategy-soil-testing-backlash/.

One of the more controversial aspects of post-fire recovery after the 2025 California wildfires was when, and by whom, the soil underlying the rubble and ash would be tested. The author looks at the role of various government agencies and officials who determine who will do the cleanup and then pay for various remediation efforts.

Brown, Julia, and Magali Gauthier. 2024. "The CZU Fire Burned More than 24K Acres across Big Basin and Nearby State Parks. Here's How They're Recovering, and Building Resilience for the Future." *Palo Alto Online*. October 21. https://www.paloaltoonline.com/outdoor-recreation/2024/11/21/the-czu-fire-burned-more-than-24k-acres-across-big-basin-and-nearby-state-parks/.

South of San Francisco where San Mateo and Santa Cruz counties meet, the 2020 CZU Fire burned nearly 90,000 acres of redwoods and parklands. It is taking years to plan and implement a park recovery program, in large part due to the bureaucratic requirements for permits, reports, and studies.

Canon, Gabrielle. 2024. "'Danger in My Backyard': Residents in a Wildfire-Prone California Town Eye More Yosemite Tourism with Unease." *The Guardian.* October 27. https://www.theguardian.com/us-news/2024/oct/27/california-wildfires-mariposa-tourism/.

Residents in the historic area surrounding Mariposa, California, not far from Yosemite are rising in opposition to a proposed Kampgrounds of America camping and glamping development they fear will bring visitors and traffic to their community, and potentially, wildfires.

Cart, Julie. 2024. "Toilet Paper and Flat Tires—the Strange Ways That Californians Ignite Wildfires." *CalMatters.* July 31. https://calmatters.org/environment/wildfires/2024/07/california-wildfires-human-causes/.

While it is well known that most wildfires are caused by humans, this article looks at the way humans actually start the fires, including some truly dumb reasons.

Cart, Julie. 2025. "California's Worsening Wildfires Explained." *CalMatters.* January 8. https://calmatters.org/explainers/california-wildfires-explained/?ref=finestra/.

The fires are getting more severe and bigger, some experts believe, because of building in and around the wildland-urban interface, or WUI. As the January 2025 fires proved, when structures become involved, the costs go up, as do the number of acres that burn more ferociously in an urban setting.

Davies, Phil. 2025. "Tourists Urged to 'Plan a Trip' to Support Los Angeles Wildfire Recovery." *Aspire.* January 21. https://www.aspiretravelclub.co.uk/news/tourists-urged-to-plan-a-trip-ti-support-los-angeles-wildfire-recovery/.

Just a few days after the start of the January 2025 wildfires in southern California, the state's tourism organization urged travelers to visit Los Angeles, telling them that more than 98 percent of the county was not in the burn area.

Derouin, Sara. 2025. "As Wildfires Rage in LA Area, Infrastructure—Including Roadways—Under Threat." American Society of Civil Engineers Source. January 15. https://www.asce.org/publications-and-news/civil-engineering-source/article/2025/01/15/as-wildfires-rage-in-la-area-infrastructure-including-roadways-under-threat/.

Civil engineers are warning state officials and the public that the state's infrastructure is at risk as a result of the January 2025 wildfires. The national organization identifies the types of damage that may not initially be assessed and considered as part of the fire's cost.

Desai, Ishani. 2024. "Health of California's Best-Known Lakes May Have Been Hurt by Back-to-Back Years of Wildfire Smoke." *The Sacramento Bee.* May 22. https://www.sacbee.com/news/local/article288639415.html.

California wildfires in 2020 and 2021 caused 70 percent of the state to be shrouded in smoke, a phenomenon that University of California, Davis researchers say has affected lakes by impacting light, temperature, and oxygen in the water.

Farber, Dan. 2025. "Wildfire Liability in California: A Primer." *Legal Planet.* April 14. https://legal-planet.org/2025/04/14/wildfire-liability-in-california-a-primer/.

Written in a way that almost everyone can understand, this article explains the sticky mess of post-fire legal actions. Who can sue whom, and why, and for how much are clearly explained with examples from recent litigation.

Garcia, Karen. 2025. "What's in the Smoke and Ash of LA-Area Fires? County Issues Health Warning." *Los Angeles Times.* January 14. https://www.latimes.com/california/story/2025-01-14/what-threats-lurk-in-the-smoke-and-ash-of-l-a-fires/.

The unique mixture of remnants of the Los Angeles fires is unlike those of forest fires, since the flames burn substances that are toxic and pose a danger to the soil, water, and air. Plastics, batteries, metals, solvents, automobiles, and furniture burn and release components generally not seen in other types of fires.

Hopkins, Sarah. 2024. "In the Park Fire, an Indigenous Cultural Fire Practitioner Sees beyond Destruction." *Inside Climate News.* September 1. https://insideclimatenews.org/news/01092024/california-park-fire-indigenous-cultural-burning/.

California's Park Fire that swept across northern California was one of the state's largest wildfires, fueled by dry grasslands. But indigenous people believe it was also an example of "good fire" that can be used to reduce the destructive potential of wildfires and a return to cultural burning.

Hubler, Shawn, and Ivan Penn. 2025. "After LA Fires, Edison Wants to Bury Power Lines in Altadena and Malibu." *New York Times*. April 11. https://nytimes.com/2025/04/11/us/la-fires-southern-california-edison.html.

Southern California Edison is using the January 2025 wildfires as justification for seeking approval to bury its power lines underground as the rebuilding process begins. The strategy is costly, but less expensive than the damage and liability of above ground power.

Katz, Lily, and Elijah de la Campa. 2024. "Thousands More People Are Moving In Than Out of Fire-and Flood-Prone America, Fueled by Migration to Texas and Florida." Redfin. August 5. https://redfin.com/news/climate-migration-real-estate-2024/.

Climate risk is now affecting migration patterns in the United States, with more people leaving or going to states such as California, Florida, and Texas where the risk is worth it.

Kluger, Jeffrey. 2025. "LA Fires Show the Reality of Living in a World with 1.5 of Warming." *Time*. https://time.com/7205644/los-angeles-california-fires-climate-change/.

While it is not totally accurate to state that California's wildfires in 2025 were due to climate change, there is little doubt that extreme weather conditions like those the state has experienced in recent years have fueled wildfires.

Lewis, James G. 2018. "Smokey Bear: From Idea to Icon." *Forest History Today*. Spring/Fall. https://foresthistory.org/periodicals/spring-fall-2018/.

This article explains in depth the story behind Smokey Bear and how the animal's image became a graphic reminder of the dangers of wildfires.

Little, Jane Braxton. 2023. "The Worst Wildfires Are Started by People." *Scientific American*. November 1. https://www.scientificamerican.com/article/the-worst-wildfires-are-started-by-people-heres-how/.

Using the example of how a father and son started the 2021 Caldor Fire in California, this article notes that 95 percent of the wildfires in the state are human-caused. There has been a gradual increase in human involvement, with this type of fire becoming more expensive because of damage to houses and the cost of fire suppression in urban areas.

Loomis, Brandon. 2025a. "Forest Thinning Aims to Curb Catastrophic Wildfires in Arizona. It Could Also Stretch Water Supplies." *Arizona Republic.* January 24. https://www.azcentral.com/story/news/local/arizona-environment/2025/01-24/srp-thins-forest-areas-in-arizona-to-save-trees-and-water-supply/.

"Letting it burn" has become as much a management strategy for contractors who fight fires as forest thinning and other methods of decreasing hazardous fuels.

MacDonald, Glen, Tamara Wall, Carolyn A. F. Enquist, Sarah R. LeRoy, John B. Bradford, David D. Breshears, Timothy Brown, Daniel Cayan, Chunyu Dong, Donald A. Falk, Erica Fleishman, Alexander Gershunov, Molly Hunter, Rachel A. Loehman, Phillip J. van Mantgem, Beth Rose Middleton, Hugh D. Safford, Mark W. Schwartz, and Valerie Trouet. 2023. "Drivers of California's Changing Wildfires: A State-of-the-Knowledge Synthesis." *International Journal of Wildland Fire* 32 (7): 1039–58. https://doi.org/10.1071/WF22155/.

The multiple authors of this research article look at the period between 1980 and 2020 to examine the increasing mean annual area burned and the severity of wildfires in the western United States and Canada.

Manke, Kara. 2024. "Researchers Reveal a Hidden Factor in California's Insurance Crisis: The 'Winner's Curse.'" *Berkeley News.* July 18. https://news.berkeley.edu/2024/-7/18/researchers-reveal-a-hidden-factor-in-californias-insurance-crisis-the-winners-curse/.

Based on a study by University of California, Berkeley researcher Meredith Fowlie, the economics phenomenon known as the "winner's curse," may be one of the under-appreciated factors that have resulted in rising home insurance prices.

McNally, Finn. 2024. "Nonprofit Questions Use of Fire Retardant during Interlaken Fire." *Leadville Herald.* July 30. https://www.leadvilleherald.com/news/article_d9f9bb36-4eaf-11ef-9ed9-275ce4f51b15.html/.

The organization Forest Service Employees for Environmental Ethics has filed suit against the Forest Service in numerous instances for the agency's failure to analyze the effects of aerial fire retardant and failing to prepare an environmental impact statement before using the chemical. This article outlines the issues and legal actions taken by the group in alleging the agency has violated its own guidelines.

Mulhollem, Jeff. 2019. "Eastern Forests Shaped More by Native Americans' Burning Than Climate Change." May 21. https://www.psu.edu/news/research/story/eastern-forests-shaped-more-native-americans-burning-climate-change/.

This article examines the work of researcher Marc Abrams, whose study of Native Americans' use of fire in the Eastern United States has resulted in the predominance of fire-adapted tree species. Abrams notes that anthropogenic fire has been the major driver of forest change within the region.

Outside, The Editors. "Forest Service Pulls $10 Million Image Campaign." *Outside.* January 7. https://www.outsideonline.com/outdoor-adventure/forest-service-pulls-10-million-image-campaign/.

A five-year public relations campaign planned by the Forest Service was withdrawn, the authors say, because the $10 million price tag had drawn criticism and backlash at a time when the agency's budget was tight and other needs were more pressing.

Pedrosa, Camila. 2024. "Night Vision, New Helicopters Are Pushing the Frontiers of Fighting Wildfires in California." *Sacramento Bee.* July 25. https://www.sacbee.com/news/california/article289610676.html/.

California's Department of Forestry and Fire Protection is considered the world's largest civil aviation firefighting unit, well known for its vast fleet of equipment and newest model of firefighting helicopters.

Perriss, James. 2023. "Phosphate-Based Fire Retardant—Saving Lives and Protecting Property for 60 Years." *International Fire Fighter.* July 25. https://iff.mag.com/?s=phosphate-based=fire-retardant/.

This article reads like a press release for the companies that supply the Forest Service with phosphate-based fire retardant for aerial firefighting operations.

Petersen, Melody. 2025. "Edison Wants to Raise Rates to Pay for Wildfires Linked to Its Equipment." January 29. https://latimes.com/environment/story/2025-01-29/edison-wants-customers-to-pay-for-wildfires-its-equipment-caused/.

Utility company Southern California Edison has asked the state Public Utility Commission to allow it to pass on to its customers the billions of dollars in damages caused by its equipment at several of the state's fires. The company has argued that it has prudently operated its power system above and beyond what regulators require.

Price, Kiley. 2024. "Fireworks Trigger Wildfires. Climate Change May Increase the Risk." *High Country News.* July 4. www.hcn.org/articles/fireworks-trigger-wildfires-climate-change-may-increase-the-risk/.

Beware July 4 fireworks, the author notes, because the holiday's events coincide with an increase in wildfires across the United States.

Pyne, Stephen J. 2017. "Vignettes of Primitive America: The Leopold Report and Fire Policy." *Forest History Today.* Spring. https://foresthistory.org/periodicals/spring-2017/.

As one of the country's premier fire historians, Pyne lends his expertise to explaining the influence of the iconic report on the development of the country's fire policies.

Reinhart, Caroline Marshall. 2024. "'Weather Whiplash' Helped Drive This Year's California Wildfires." *Inside Climate News.* September 12. https://insideclimatenews.org/nws/12092024/california-wildfires-weather-whiplash/.

Written before the January 2025 wildfires in southern California, the article notes that heavy winter rains in 2023 and 2024, coupled with two years of drought in 2020 to 2022, were responsible for producing the growth in grasses and shrubs that were fueling wildfires throughout the state.

Rodd, Scott. 2023. "Wildfire Victims Left 'In the Dark' after US Forest Service Briefs Congress about the Caldor Fire." CapRadio. March 20. https://capradio.org/articles/2023/03/20/wildfire-victims-left-in-the-dark-after-us-forest-service-briefs-congress-about-the-caldor-fire/.

In 2021, the Caldor Fire between Sacramento and Lake Tahoe grew to become one of the state's largest wildfires. The Forest Service has been heavily criticized for its failure to act on wildfire risks that had been identified decades prior to the fire.

Saam, Kelli. 2025. "Lawsuit Aims to Stop Largest-Ever Fire Prevention Project on Plumas Forest." *Action News Now.* January 15. https://www.actionnewsnow.com/news/lawsuit-aims-to-stop-largest-ever-fire-prevention-project-on-plumas-nat-forest-article/article/.

The environmental group Feather River Action filed a lawsuit in 2025 seeking an environmental impact statement on the $650 million, 200,000-acre Community Protection Project on the Plumas National Forest in California.

The project was initially fast tracked under an emergency declaration, which the group believes failed to look at the impact of logging, burning, and herbicides.

Sadasivam, Naveena. 2024. "The American West's Megafires Are Silencing Birds." *Grist*. August 14. https://grist.org/extreme-weather-wildfire-birds-sound-study/.

Smoke from wildfires may be one of the factors that is causing birds to become quieter, according to a UCLA study that examines wildlife sensitivity to air pollution. The study looks at the soundscapes of rural Washington using acoustic monitors, an indicator that is the subject of controversy among researchers.

Schoennagel, Tania, Cara R. Nelson, David M. Theobald, and Teresa B. Chapman. 2009. "Implementation of National Fire Plan Treatments Near the Wildland-Urban Interface in the Western United States." *Proceedings of the National Academy of Sciences.* June. https://www.pnas.org/doi/abs/10.1073/pnas.0900991106/.

Written for a scholarly audience, this report is one of the few that carefully examines whether fuel treatments are effective when used in specific settings such as the wildland-urban interface.

Smith, Hayley. 2024. "Millions of Californians Live Near Oil and Gas Wells That Are in the Path of Wildfires." *Los Angeles Times.* July 21. https://www.latimes.com.environment/story/2024-07-21/wildfires-pose-increasing-risk-to-california-oil-wells/.

As if wildfires themselves were not a problem, a University of California, Berkeley researcher is now warning residents that there are an estimated 100,000 oil and gas wells in nineteen states west of the Mississippi River that are in areas that either have burned in recent decades and/or face a high risk of burning in the future.

Stringer, Grant. 2024. "In Oregon, a New Program Is Training Burn Bosses to Help Put More 'Good Fire' on the Ground." *Inside Climate News.* January 29. https://insideclimatenews.org.news/29012024/oregon-training-burn-managers/.

The importance of training firefighting personnel how to use "good" fire, or prescribed burning, is now becoming a key factor in leading states and private landowners to reçruit and train professionals to ignite and manage these types of planned fires.

Sumagaysay, Levi. 2025. "California Homeowners Will Have to Fund Half of High-Risk Insurer's $1 Billion 'Bailout.'" *CalMatters*. February 11. https://calmatters.org/economy/2025/02/homeowners-insurance-costs-rising-in-california-fair-plan/.

Just a month after the 2025 California wildfires, the state's Insurance Department approved a request that allowed the state's FAIR Plan to impose a $1 billion charge on insurance companies, as the author explains.

Tibbitts, Dana. 2024. "License to Burn: Wildfire as the Ultimate Private-Public Partnership." *The Nevada Globe*. April 22. https://thenevadaglbe.com/articles/license-to-burn-wildfire-as-the-utlimate-private-public-partnership/.

Public-private partnerships are becoming more prevalent as various memorandums of understanding are being developed to "treat" forests and rangelands. These agreements are becoming more controversial, especially in the Tahoe Basin, as this article explains.

Toohey, Grace. 2024. "Hillside Erosion Worsening in California due to Wildfires and Intense Rain." *Los Angeles Times*. August 31. https://www.latimes.com/environment/story/2024-08-31/erosion-following-california-wildfires-is-getting-worse-water-infrastructure-ecosystems-could/.

Based on a US Geological Survey study, California has experienced increasing erosion after wildfires over the last three decades. The resulting post-fire hillside erosion has affected water resources and ecosystems, and is likely to worsen as a result of climate change.

Toohey, Grace, Alex Wigglesworth, and Rong-Gong Lin II. 2025. "Tiny Burning Embers Flew Miles, Causing LA Fire Destruction on Historic Scale." *Los Angeles Times*. January 9. https://latimes.com/california/story/2025-01-09/burning-embers-flew-miles-brought-mass-destruction-across-l-a-county/.

Firefighters often remind homeowners that it is not the walls of flames that burn houses, but the embers from those fires that blow across long distances. The erratic Santa Ana winds that blew through Los Angeles in January 2025 spun embers an estimated 2 and 3 miles ahead of the established blazes.

Udasin, Sharon. 2024. "Wildfires Are Becoming Faster and More Furious in the US West." *The Hill*. October 24. https://thehill.com/policy/equilibrium-sustainability/4951845-wildfires-faster-us-west/.

The term "fast fires" is now being used to describe fast-growing flames which the author notes are now responsible for nearly 90 percent of wildfire-related damage in recent years.

Whelton, Andrew J. 2024. "Wildfires Don't Just Burn Farmland—They Can Contaminate the Water Farmers Use to Irrigate Crops and Support Livestock." *The Conversation.* August 13. https://theconversation.com./wildfires-dont-just-burn-farmland-they-can-contaminate-the-water-farmers-use-to-irrigate-crops-and-support-livestock/.

In this report on the August 2023 wildfires across Maui, the author worked with the Hawaii Department of Agriculture to assess the state's damaged water systems, an expensive part of rural infrastructure that is rarely considered.

Whipple, Dan. 1993. "Nature Turns Massive Fire at Yellowstone to Its Benefit." *Los Angeles Times.* October 12. https://articles.latimes.com/1993-10-13/news/mn-45340_1_greater-yellowstone-area/.

The author is a respected reporter whose coverage of the Yellowstone Fires includes several follow-up articles that document the impact of the blazes on wildlife, local ecological resilience, and future wildfire policies.

Whipple, Dan. 2015. "Yellowstone Ablaze: The Fires of 1988." Wyohistory.org. June 27. www.wyohistory.org/encyclopedia/Yellowstone-ablaze-fires-1988/.

This concise coverage of the Yellowstone blazes covers themes seldom addressed by others, including how the park had recovered five years later, media coverage that was often contradictory, and the debate over the "Let it burn" policy.

Wigglesworth, Alex. 2024. "Wildfire Retardant Is Laden with Toxic Metals, USC Study Finds." *Los Angeles Times.* November 16. https://www.latimes.com/environment/story/2024-11-16/popular-fire-retardant-has-high-levels-of-heavy-metals-usc-study-finds/.

Although firefighters say there is no substitute for using fire retardant, a study conducted by researchers at the University of Southern California found that one popular type contains toxic metals that endanger ecosystems.

Williams, Gerald W. 2000. "Introduction to Aboriginal Fire Use in North America." *Fire Management Today* 60 (3): 8–12. https://www.frames.gov/documents/usfs/fmt/fmt_60-3.pdf.

The terms indigenous and native and aboriginal are often used interchangeably, and Williams uses the latter to emphasize the primitive nature of fire use.

Winerup, Michael. 2013. "Lessons from the Yellowstone Fires of 1988." *New York Times*. September 2. https://www/nytimes.com/2013/09/02/booming/lessons-from-the-yellowstone-fires-of-1988.html.

On the 25th anniversary of the Yellowstone fires that burned nearly one-third of the national park, the author describes how the nation's prescribed fire and burn policies have not completely changed as a result.

Wolfe, Daniel, and Aaron Steckelberg. 2024. "A Surprising Byproduct of Wildfires: Contaminated Drinking Water." *Washington Post*. July 29. https://www.washingtonpost.com/climate-environment/interactive/2024/wildfire-drinking-water-supply-danger/.

Called a "hidden threat," the contamination of drinking water sources by wildfires that occur in forested areas can lead to increased risk from cancer-causing and toxic substances.

Government Documents and Other Monographs

Copernicus Climate Change Service. n.d. "Global Climate Highlights 2024." https://climate.copernicus.edu/global-climate-highlights-2024/.

The message from the organization included in this annual report is clear: that 2024 was the warmest year for global temperatures in records dating back to 1850. The document cites other major findings that show the last ten years have been the warmest ten years on record.

National Interagency Fire Center. 2025. *Statistics*. https://www.nifc.gov/fire-information/statistics/.

Although several programs within this agency have been threatened with budget cuts, this is still the best source of up-to-date information, and especially longitudinal data.

National Oceanic and Atmospheric Administration. 2023. *National Weather Service: Fire Weather*. https://www.weather.gov/fire/.

There is a fire weather "community" that use models and technology to support decisions involving wildland fire. This document explains the use of forecasting and decision-support tools to identify fire weather environments.

National Oceanic and Atmospheric Administration. n.d. "Wildfire Climate Connection." https://www.noaa.gov/noaa-wildfire-climate-connection/.

The agency cites changes in climate that create warmer, drier conditions that lead to longer and more active fire seasons.

National Park Service. 2024. "Planning for Resilience: Understanding the Connections between Forests, Fire, Water, and Drought." https://home.nps.gov/articles/planning-for-resililence-understanding-the-connections-between-forests-fire-water-and-drought.htm.

This report, developed by the Desert Research Institute and Yosemite National Park, takes a holistic view of the landscape to determine which areas would benefit from low-intensity fires.

New Mexico. Energy, Minerals, and Natural Resources Department. n.d. "History of the Cooperative Forest Fire Prevention Campaign." https://www.emnrd.nm.gov/sfd/smokey-bear-historical-park/history-of-the-cooperative-forest-fire-preventiion-campaign/.

Smokey Bear began his journey in New Mexico when the cub was found after a fire, and this document explains some of that journey and its importance in the development of fire prevention campaigns.

Organisation for Economic Cooperation and Development. 2023. *Taming Wildfires in the Context of Climate Change: The Case of the United States.* Paris: OECD Publishing. https://doi.org/10.1787/dd00c367-en.

This international agency uses the United States as a case study of the nexus between growing concerns about climate change and wildfires, and how the federal government is trying new strategies to reduce severity.

Sedlar, Matt. 2024. "US Forest Service Decision to Halt Prescribed Burns in California Is History Repeating." Center for Economic and Policy Research. October 28. https://cepr.net/us-forest-service-decision-to-halt-prescribed-burns-in-california-is-history-repeating/.

This brief article explains the October 2024 decision by the US Forest Service to halt prescribed burns "for the foreseeable future" as a precautionary measure in case major wildfires should erupt.

US Forest Service. Department of Agriculture. 2022. *Confronting the Wildfire Crisis: A Strategy for Protecting Communities and Improving Resilience in America's Forests.* https://www.fs.usda.gov/sites/default//files/confronting-wildfire-crisis.pdf/.

Stating the obvious—that wildfires are getting bigger and more severe—this generic document calls for strategies that combine partnerships and a four-fold increase in fuels treatment.

US Forest Service. Department of Agriculture. 2023. *Wildfire Hazard Potential.* https://www.firelab.org/project/wildfire-hazard-potential/.

Updated in 2025, this document summarizes maps of conditions in the landscape at the end of 2020. As its name implies, the modeling of conditions helps planners determine where wildfires are most likely to erupt in upcoming fire seasons.

US Forest Service. Department of Agriculture. 2023. *Wildfire Risk to Communities.* https://wildfirerisk.org/.

This is a relatively user-friendly website that is designed to help stakeholders understand what wildfire risks and resources are available in their own community.

Western Forestry Leadership Coalition. 2022. *The True Cost of Wildfire in the Western US.* Edgewater, CO: Western Forestry Leadership Coalition. www.thewflc.org/sites/default/files/TrueCostofWildfire.pdf.

The organization argues that it is difficult to ascertain the cost of a wildfire, sometimes because there are a multiplicity of jurisdictions involved. Costs include suppression as well as damage to natural resources, structures and property, infrastructure, and difficult to measure items like the loss of life, health, and well-being.

Wildland Fire Leadership Council. 2023. *National Cohesive Wildland Fire Management Strategy: Addendum Update.* https://forestsandrangelands.gov/documents/documents/strategy-national-cohesive-wildland-fire-mgmt-strategy-addendum-update-2023.pdf.

This 2023 document is a supplement to the 2014 document the group produced to address four topics: climate change, workforce capacity, health and well-being, community resilience, and diversity, inclusion, and environmental justice.

Williams, Gerald W. 2005. *The USDA Forest Service: The First Century.* Washington, DC: US Department of Agriculture. Forest Service. https://www.fs.usda.gov/sites/default/files/media/2015/06/The_USDA_Forest_Service_TheFirstCentury.pdf.

This historical account of the agency is by far the best, although lengthy, in providing detailed information on its formation and growth.

Electronic Media

Agee, James. 2022. "Fire Ecology, Why We Need More Fire." WildFed Podcast #146.

https://www.wild-fed.com/podcast/146.

As one of the nation's top experts on forest ecology and wildfires, Agee argues that fire is an essential element of our landscape. But, he says, there has been a resistance to use fire due to cultural issues related to landownership that have kept fire out of the ecological toolbox.

The Big Burn. 2015. The American Experience. PBS Videos. https://www.pbs.org/video/american-experience-big-burn/.

The Great Fire of 1910, known as The Big Burn, raged through the Rockies and led to a wave of expansion in the country's firefighting services and policies.

Dick, Vester. 1984. *Up in Flames: A History of Fire Fighting in the Forest.* Forest History Society.

Archival footage of some of the country's major wildfires, along with examples of the development of fire technology like aerial tankers, are illustrated in this short (29 minutes) film by the noted filmmaker.

Elemental: Reimagine Wildfire. 2023. Balance Media and Go Alliance. https://www.elementalfilm.com/.

Using the town of Paradise, California, as a backdrop, the video includes commentary from top wildfire experts and examples from researchers to help us understand why some structures burned and others did not. Coverage of indigenous fire practices used today shows how fire can benefit the environment and make communities safer.

Extreme Wildfire: Deadly, but Not for the Climate. 2021. PBS Digital Studios. https://www.pbs.org/video/extreme-wildfire-deadly-but-not-for-the-climate-480bae/.

Fire tornadoes like the one that started the Carr Fire in 2018 are becoming more prevalent as climate change affects the destructive forces of wildfires.

Fire and Ash: Devastation in L.A. 2025. NBC News. https://www.youtube.com/watch?v=hBPkRIAoers.

Dateline host Lester Holt narrates this documentary coverage of the first four days of the January 2025 wildfires using interviews and coverage from NBC News as the backdrop.

Fire in Paradise. 2021. Frontline. PBS Videos. https://www.pbs.org/wgbh/frontline/documentary/fire-in-paradise/.

The Camp Fire is not only one of the expensive fires in US history, it was also one of the most devastating in how it destroyed entire Northern California communities. This Frontline episode looks at what it calls the "inside" story of the fire, including why it happened, the firefighters, and the victims.

How the LA Wildfires Unfolded and What's Needed for Recovery. 2025. BBC News. https://www.youtube.com/watch?v=nBZs2_hezdY/.

This BBC presentation focuses on the human side of the wildfires, with heartwarming interviews of the victims of the fire and how they intend to return.

How the US Spends Billions Fighting Wildfires. 2025. Business Insider. https://www.youtube.com/watch?v=slNxj010iTs/.

The costly 2025 wildfires brought attention to innovative methods of dealing with wildfires, from the $30 million Super Scooper aircraft that spreads water over wildfires to new techniques that are being used to fireproof homes.

L.A. Fires Expose Long-Standing Local and National Water Infrastructure Challenges. 2025. Brookings Institution. Metro Blueprint. https://www.

youtube.com/watch?v=gPyko3Qzvqs&list=PLrI3Iy0dqgpR8pI0vvq-4BZIyP9C0twmw&index=1/.

Experts from UCLA and the Brookings Institution note that the January 2025 fires in Los Angeles were different from any of those analyzed since about 2017.

LA Firestorm: The Devastation and Determination of Los Angeles. 2025. Spectrum News. January 17. https://www.youtube.com/watch?v=pV07XWni-B0.

Hosted by Spectrum News' Amrit Singh, this video takes viewers through the region's destruction and devastation. It also, however, sheds light on the determination of the people of Los Angeles.

Operation Earth: Wildfires. 2020. National Aeronautics and Space Administration. https://science.nasa.gov/resource/video-operation-earth-wildfires/.

NASA satellites provide a unique perspective on the tools the agency has to predict fire behavior, monitor active fires, and deal with post-fire recovery. The film uses the 2017 Complex Fire in Northern California to illustrate how NASA provides assistance to on-the-ground responders.

Post Fires, the LA Times Is Digging Up Dirt. 2025. Boiling Point Podcast. https://open.spotify.com/episode/37liDiNMMkUmwGG6DM9sEG?go=1&sp_cid=169a4819d4fc872c8e419db0b1ef1223&utm_source=embed_player_p&utm_medium=desktop&nd=1&dlsi=0e8c5b69b6c242b7/.

In this half hour episode, the focus is the results of the series of soil tests conducted by reporters for the *Los Angeles Times* after the January 2025 wildfires. What they have uncovered runs counter to what federal officials were saying to the public about the toxicity of the remaining dirt where their houses once stood.

The Real Story of the LA Firestorm. 2025. PBS Terra. https://www.youtube.com/watch?v=PiX9t_wovEY.

PBS is on the ground in Los Angeles on Day 1 of the LA fires. The video includes fire officials and leading scientists to learn what went wrong in LA, and the one thing that could stop these unstoppable fires.

Schultz, Tom. 2025. *Chat with the Chief.* March 13. US Forest Service. https://www.fs.usda.gov/inside-fs/leadership/now-available-chat-chief-episode-1/.

The Chief of the US Forest Service started this new monthly series of interviews with employees just after he was named to head the agency. Employees are asked to submit questions in advance. The first episode features Martina Barnes, a district ranger on the Green Mountain and Finger Lakes National Forests.

Weathered: Inside the LA Firestorm. 2025. PBS Videos. https://www.pbs.org/video/weathered-inside-the-la-firestorm-l31r0b/.

Host Maiya May looks at a different side of the 2025 fires by examining the causes of the disastrous blazes. The goal is to find ways to prevent future catastrophic infernos.

The West Is Burning. 2021. PBS Videos. https://www.pbs.org/video/the-west-is-burning-P2u7ow/.

In this somewhat controversial film about western region forests, the argument is made that current management practices and litigation are responsible for the growing number and scale of catastrophic fires. The solution is said to be an interdisciplinary approach to ecological restoration and cooperation among stakeholders.

Wildfires. 2025. Energy Switch. PBS Videos. https://www.pbs.org/video/wildfires-d2azoy/.

This episode is part of a series, and it focuses on how some areas are having more wildfires, while others are experiencing few numbers. Part of the reason is that more Americans are living in fire-prone areas called the wildland-urban interface.

Other Sources

Mindock, Clark. 2023. "Wildfire Retardant Violates US Water Law, Judge Says." *Reuters.* May 26. https://www.reuters.com/legal/government/wildfire-retardant-violates-us-water-law-judge-says-2023-05-26/.

Peterson, Molly, Lisa Pickoff-White, April Dembosky, and Danielle Venton. 2020. "Even after Care Homes Abandoned Residents, California Still Isn't Ready for Wildfires." KQED. August 10. https://www.kqed.org/science/1968076/even-after-care-homes-abandonded-Residents-california-still-isnt-ready-for-wildfires/.

Peterson, Paula. 2024. "Technology Being Used as a Partner in Forest Fuels Treatment at Lake Tahoe." SouthTahoeNow.com. May 2. https://southtahoenow.com/05/02/2024/technology-being-used-partner-forest-fuels-treatment-lake-tahoe/.

Rodd, Scott. 2023. "Caldor Fire Survivors Are Taking Legal Action against US Forest Service." CapRadio. August 2. https://www.capradio.org/articles/2023/08/02/caldor-fire-survivors-are-taking-legal-action-against-us-forest-service/.

Roos, Dave. 2023. "Native Americans Used Fire to Protect and Cultivate Land; Indigenous People Routinely Burned Land to Drive Prey, Clear Underbrush, and Provide Pastures." The History Channel. August 11. https://www.history.com/articles/native-american-wildfires/.

Western Fire Chiefs Association. 2024. *Wildland Urban (WUI) & Firefighter Challenges.* September 17. https://wfca.com/wildfire-articles/wildland-urban-interface-wui-firefighter-challenges/.

Wibbenmeyer, Matthew, and Annie M. Tastet. 2021. *Wildfires in the United States 101: Context and Consequences.* New York: Resources for the Future. https://www.rff.org/publications/explainers/wildfires-in-the-united-states-101-context-and-consequences/.

Wibbenmeyer, Matthew, and Lauren Dunlap. 2022. *Wildfires in the United States 102: Policy and Solutions.* New York: Resources for the Future. https://www.rff.org/publications/explainers/wildfires-in-the-united-states-102-policy-and-solutions/.

7

Chronology

October 8, 1871: Wisconsin's Peshtigo Fire kills at least 1,150 people and burns 1.2 million acres during one of the driest summer and fall seasons on record in the Midwest. Logging debris, fueled by 100-mph winds, form a storm front that burns over 1,000 structures.

September 5, 1881: At 1 million acres destroyed, Michigan's Thumb Fire results in the deaths of at least 282 people. The fire is attributed to drought, high winds, and logging practices, partially obscuring the sun along the East Coast because of soot and ash. The damage estimate from the wildfire is $2.3 million, equivalent to $76.5 million when adjusted for inflation.

September 24, 1889: The Santiago Canyon Fire, bolstered by Santa Ana winds, burns 300,000 acres in southern California, becoming the largest fire in the state's history before wildfires were officially recorded. The fire starts in a sheepherder's camp, killing thousands of sheep and destroying large quantities of grain.

The New Jersey Army National Guard assists local wildfire fighting efforts using water buckets and a Black Hawk helicopter to extinguish flames in Ocean County, New Jersey, April 28, 2025. The Jones Road Fire, caused by arson involving young persons in the Forked River Mountains Wilderness Area, burned 15,300 acres (US Army National Guard photo by Sgt. Michael Schwenk).

September 1, 1894: Minnesota's Great Hinckley Fire kills 418 people, although historians believe hundreds of Native Americans were also killed. The fire burns more than a thousand structures and an estimated 250,000 acres in just four hours. The cause of the fire is a prolonged drought, high temperatures, and several small fires that merged from piles of logging debris.

September 8–12, 1902: The Yacolt Burn, named for multiple fires in Washington and Oregon, leads to the death of sixty-five persons and burns about a million acres due to drier than normal summer weather and a buildup of logging slash.

February 1, 1905: Management of the forest reserves shifts to the US Forest Service (USFS) through the Transfer Act. The agency takes on responsibility for developing professional firefighting standards, hiring rangers, and hiring local men to help put out fires.

August 20–21, 1910: A complex of fires called the Great Fire of 1910 (also called the Big Burn, or the Big Blow Up) destroys 3 million acres of mostly virgin forest in northern Idaho, eastern Washington, and western Montana. About $1 billion worth of timber is destroyed, and eighty-seven people, of whom seventy-eight were firefighters, are killed.

March 1, 1911: Congress enacts the Weeks Act expanding the USFS and establishes federal-state programs for fire control. The agency also begins a program of fire research.

1914: California District Forester Coert DuBois leads research on light burning and publishes *Systematic Fire Protection in California.*

October 12, 1918: Minnesota and Wisconsin's Cloquet and Moose Lake Fires lead to second deadliest fires in US history, with 453 fatalities. The fires are caused by sparks on rail lines from passing trains during the driest

	October in nearly five decades, destroying thirty-eight communities.
November 14, 1921:	The first-ever National Fire Control Conference, also known as the Mather Field Conference, is organized by Chief Forester Bill Greeley to review fire policy, techniques, and lexicon to standardize USFS policy and operations.
June 7, 1924:	The Clark-McNary Act expands cooperative programs for state and federal fire agencies, allowing the USFS to administer grants equal to the amounts contributed to firefighting by states.
1926:	USFS adopts a policy that would allow areas of 10 acres or less to burn, requiring the suppression of all fires over 10 acres.
May 22, 1928:	The McSweeney-McNary Act officially authorizes fire research, and assigns the USFS responsibility for conducting and administering research programs.
September 7–20, 1932:	Matilija Fire in Ventura County, California, destroys 220,000 acres of rugged terrain, mostly in what is now the Los Padres National Forest, that had not been burned in several decades. Supplies had to be carried to the area on mules or dropped from airplanes.
1935:	Participants at the annual meeting of the Society of American Foresters argue that wildfires are useful for forest ecosystems.
1935:	Chief Forester Ferdinand Silcox announces an "experiment on a continent scale" policy that stipulates fires must be controlled by 10 a.m. the day following the report of a fire, or if that is not possible, controlled by 10 a.m. the following day.
1939:	Two new Forest Service programs are implemented: smokejumpers who parachute into the field to control

snag fires in remote area; and shock troop crews of forty men trained to fight large fires.

1939: A poster of a forest ranger in the image of Uncle Sam pointing to a raging forest fire is presented to the public with the caption, "Your Forest—Your Fault."

July 12, 1940: The first smoke jumping on a forest fire takes place on the Martin Fire on the Nez Perce National Forest in Idaho.

1942: USFS organizes the Cooperative Forest Fire Prevention Program.

1942: Raymond Conarro coins the term "prescribed burning" to describe scientifically informed fire practices controlled by humans.

1943: USFS adopts policy of using burning for fuel reduction in the southern United States.

1944: The characters in the 1942 Walt Disney film *Bambi* are loaned to the Cooperative Forest Fire Prevention Program for a poster promoting fire prevention, but only for one year.

August 9, 1944: A bear is chosen as the symbol for forest fire prevention, becoming the longest-running public service advertising campaign in US history.

1947: Smokey Bear's fire prevention message, "Remember, only YOU can prevent forest fires" is coined.

1947: The USFS tests methods of detecting and suppressing fire using aerial resources in the northern Rockies.

1948: The USFS establishes the Division of Fire Research to develop a national fire research agenda supervised by foresters with experience in engineering and economics.

August 5, 1949: Thirteen smokejumpers die in a fire in Montana's Helena National Forest in the Mann Gulch Fire when flames flare up in high winds.

May 9, 1950: An orphaned bear is found after a wildfire in the mountains of Capitan, New Mexico. The bear is initially named Hotfoot Teddy, but later is renamed Smokey after the bear in the poster. The bear is sent to the National Zoo in Washington, DC, and used as a symbol of fire prevention.

1953: A Junior Ranger Program is established to teach children how to help Smokey Bear prevent forest fires.

1954: Operation Firestop is created as a one-year-long crash program to combine the resources of several agencies, some using military-style tactics and equipment, to study new technologies for fighting fires.

1961: The Inter-Regional Fire Suppression (IRFS) program is developed, establishing six thirty-man rapid deployment crews across the Western United States. The crews are stationed near airports for quick transportation to high-priority fires.

November 5–8, 1961: The Bel-Air Fire in southern California becomes one of the biggest wildfire events covered by television stations, destroying 505 luxury homes and 6,090 acres. The fire receives widespread public attention in part due to the homes of celebrities that burn. The fire results in the banning of wood shingle roofs and stringent brush clearing policies.

March 4, 1963: The Leopold Report, written by Aldo Leopold's oldest son, wildlife biologist Starker Leopold, recommends the controlled use of wildfire as one easy-to-apply and inexpensive way of managing vegetation to improve wildlife habitat.

1968:	The National Parks Service recommends that prescribed burning be restored as its policy in NPS units.
1969:	The Interagency Fire Center is established in Boise, Idaho, to consolidate the national response to wildfires.
September 22–October 4, 1970:	The Laguna Fire in eastern San Diego County, which started with downed fire lines during a period of high Santa Ana winds, is, at the time, the second-largest fire in California history. The wildfire burns 175,425 acres, destroying 1,382 structures. An estimated eight persons were killed.
1971:	The USFS revises its policy to allow some lightning fires to burn as natural prescribed fires in wilderness areas.
October 22, 1976:	Enactment of the National Forest Management Act redefines the mission of the USFS, reducing the agency's role in fighting wildfires.
1976:	The National Wildfire Coordinating Group is established to standardize training, equipment, and certification for fire agencies.
1978:	USFS abandons its own policy that all fires were to be extinguished by 10 a.m. the following day, shifting to managing fire suppression costs so that they are consistent with land and resource management strategies.
June 14–November 18, 1988:	Fires burn 45 percent of the acreage in Yellowstone National Park at a cost of about $3 million per day. The fires are spread by drought conditions and high winds, burning for several months before being extinguished by cooler weather. Some ignitions were allowed to burn as prescribed natural fires as

a resource benefit in what becomes a controversial management approach.

June 25–26, 1990: Six firefighting inmates from the Perryville State Prison in Phoenix are killed during the Dude Fire near Payson, Arizona.

October 19, 1991: The Tunnel Fire in the Oakland and Berkeley, California Hills destroys 2,843 homes, marking what is considered to be the modern era of urban wildfires. The fire kills twenty-five people.

July 6, 1994: The lightning-caused Storm King Fire (also known as the South Canyon Fire) kills twelve wildland firefighters, members of the Prineville, Oregon Hotshots crew, and two helitack firefighters near Glenwood Springs, Colorado.

1997: Chief Mike Dombeck alters the USFS focus on ecosystem management through the Natural Resource Agenda, emphasizing watershed health and restoration; sustainable forest management; national forest roads; and recreation.

1998: The Joint Fire Science Program is instituted under the aegis of both the Department of the Interior and the USFS to consolidate federal fire research.

May 4–July 20, 2000: The Cerro Grande Fire in New Mexico begins as a backfire to contain a prescribed fire, and threatens the Los Alamos National Laboratory.

2001: The Smokey Bear slogan gets an update in response to fires in non-forested areas, becoming "Only You Can Prevent Wildfires."

June 8–July 18, 2002: An employee burning a letter causes the Hayman Fire in Colorado, spreading over four counties, and damaging 600 structures, at a cost of $39 million.

June 2002:	Two fires merge in east-central Arizona, forming the Rodeo-Chedeski Fire that destroys 462 structures, with a suppression cost of about $153 million. Investigations later find the fire was started by a man seeking work as a firefighter.
July 12–December 31, 2002:	The lightning-caused Biscuit Fire in Oregon's Siskiyou National Forest burns a half million acres at a cost of $150 million.
October 25–November 5, 2003:	The Cedar Fire in San Diego becomes part of a complex of fires, initially caused by a lost hunter's signal fire, burning 280,000 acres and leading to fourteen deaths.
December 3, 2003:	President George W. Bush signs the Healthy Forests Restoration Act, in part directing the Secretary of Agriculture to develop a program to reduce hazardous fuels on federal land to reduce fire frequency.
October 21–22, 2007:	Arcing power lines and heavy winds started a small fire in the Witch Creek area of eastern San Diego, eventually making its way to the San Diego city limits. The next day, the Guejito Fire starts in the San Pasqual Valley, and the two fires merge, burning nearly 198,000 acres. Countywide, 1,141 residences were destroyed, with two deaths. San Diego Gas & Electric Company pays over $2 billion in settlements without admitting liability.
2008:	California implements building codes that mandate that new construction in high fire hazard areas be built with ignition-resistant roofs, decks, and eaves.
June 28–July 9, 2013:	Started by a lightning strike, the Yarnell Hill Fire near Prescott, Arizona covers 8,000 acres and kills nineteen firefighters unable to find shelter when the fire changes its behavior.

2014: The Wildland Fire Leadership Council establishes the National Cohesive Wildfire Management Strategy setting goals and activities to reduce wildfire risk.

November 23–December 22, 2016: The Great Smoky Mountain Wildfires devastate 17,900 acres near Gatlinburg and Pigeon Forge, Tennessee, along with parts of the Great Smokey Mountains National Park, claiming fourteen lives.

October 8–28, 2017: A series of more than 250 fires, collectively named the 2017 October Fire Siege, converge on California, killing forty-four people. One blaze, the Atlas Fire, burns 51,624 in California wine country near Napa, destroying 783 structures and claiming six lives, with a monetary loss of $3 billion.

October 8–31, 2017: The Tubbs Fire sweeps through the city of Santa Rosa, California, destroying 5,643 structures at a dollar loss of $11 billion, with twenty-two lives lost.

November 8–25, 2018: A downed electrical power line results in the Camp Fire in northern California, killing eighty-six people and destroying nearly 19,000 structures in the small communities of Paradise and nearby Concow, California. The total dollar cost is estimated at $12.4 billion. That same day, southern California's Woolsey Fire begins in southeast Ventura County and Malibu. Nearly 100,000 acres burn, along with 1,643 structures, with a dollar loss of $5.2 billion. Three people are killed.

January 29, 2019: PG & E Corporation and Pacific Gas and Electric Company jointly file for bankruptcy protection due to losses following several wildfires in northern California between 2015 and 2018.

2019: The National Park Service launches its Wildland Fire Strategic Plan outlining its approach to wildfire prevention, impact mitigation, response, and restoration efforts on 84 million acres of land under its authority.

July 1, 2020: California's Fire Victim Trust is established to evaluate, administer, process, and resolve claims arising from twenty-two wildfires in northern California. The trust provides compensation for economic and non-economic damages caused by the fires.

August 16–17, 2020: The August Complex Fire begins as a series of thirty-eight fires ignited by lightning strikes along California's Coast Range. By the time it is extinguished on November 12, the fire has burned 1,032,648 acres, destroying 935 structures and killing one firefighter.

September 7–8, 2020: Five megafires over the Labor Day Weekend in Oregon burn nearly 850.000 acres, killing eleven persons, and destroying more than 4,000 structures.

July 13, 2021: The Dixie Fire burns across five California counties, burning 963,309 acres and destroying more than 1,300 structures. It became one of US history's most expensive wildfires to suppress, costing $637.4 million. The wildfire is caused by a tree that hits electrical distribution lines owned by Pacific Gas & Electric Company.

August 2021: The Los Angeles City Council directs the fire department to take stock of its readiness to fight an outbreak of wildfires.

December 30, 2021: The most destructive fire in Colorado history, the Marshall Fire, starts out as a grass fire in Boulder County. Over 37,000 people are evacuated, two are killed, and over 1,000 structures are destroyed.

2022: The USFS releases a wildfire risk management strategy that establishes enhanced fuel management as a main

	policy objective. The strategy calls for the agency to treat 20 million acres across federal lands.
April 11, 2022:	Pacific Gas & Electric agrees to pay more than $55 million to avoid criminal prosecution in the 2021 Dixie Fire and the 2019 Kincade Fire in California. The settlement agreement accelerates payments to hundreds of people whose homes were destroyed in wildfires ignited by PG & E's equipment.
2023:	The Wildland Fire Leadership Council publishes the National Cohesive Wildland Fire Strategy update.
January 2023:	The Los Angeles Fire Chief asks the city for $7 million to assemble its own specialized wildland unit to respond to and contain wildfires rather than relying on hand crews from other jurisdictions. The lack of a professional hotshot crew is called a significant area of weakness in the city's ability to contain wildfires. The city responds in May 2023 by approving $519,000 for a new unit, about 7 percent of the fire department's request.
June 2023:	The USFS releases the agency's National Prescribed Fire Resource Mobilization Strategy to prioritize prescribed fire on a larger geographic scale, including the expansion of prescribed burning on national forest lands.
August 8–9, 2023:	The historic town of Lahaina, Hawai'i, suffers a devastating fire resulting in 102 confirmed fatalities and widespread destruction of homes and businesses after winds down powerlines in a grassy area. The estimated dollar loss of $6 billion includes about 3,000 structures, even though just over 2,100 acres burn.
February 26–March 16, 2024:	Texas' Smokehouse Creek Fire burns 1.1 million acres, one of only ten US wildfires to burn 1 million acres or more. Estimates of damage are $6.2 million because most of the fire occurred in sparsely populated rural land.

February 26, 2024: The Windy Deuce Fire in Potter County, in the Texas Panhandle, burns over 140,000 acres. The cause of the fire is a downed utility pole.

November 8, 2024: Wildfires, from the 2-acre Prospect Park Fire in a densely vegetated part of Brooklyn to the Jennings Creek Fire that burned more than 5,300 acres along the New York-New Jersey border, send smoke across the New York skyline. The highly publicized fires are due largely to an extended drought, with New Jersey and Delaware reporting their driest October weather on record. An eighteen-year-old volunteer firefighter is killed battling the Jennings Creek Fire.

January 7–31, 2025: A series of wildfires in southern California start in Pacific Palisades and Altadena, spreading to Malibu and Pasadena, destroying more than 16,200 structures. The blazes are a perfect storm driven by hurricane-force wild gusts, unseasonably dry conditions, and multiple fires breaking out one after another in the same geographic area. The Eaton Fire burns 14,117 acres with sixteen deaths reported.

February 23, 2025: California Governor Gavin Newsom asks Congress to approve nearly $40 billion to help the Los Angeles area recover from deadly wildfires.

February 25–28, 2025: Dismal River Ranch Fire in McPherson and Hook Counties, Nebraska burns 50,000 acres of grassland vegetation.

March 14, 2025: Wildfires break out in Texas and Oklahoma due to dry vegetation and high winds, scorching grasslands in the Panhandle area. The 33,000-acre 33 Road Fire in Logan and Payne Counties, Oklahoma destroys fifty structures, while the Stillwater Fire destroys at least seventy-four structures.

April 21, 2025: Jones Road Wildfire, caused by arson involving young persons in the Forked River Mountains Wilderness Area, burns 15,300 acres, destroying one commercial building and other outbuildings and vehicles.

July 4, 2025: Lightning-caused Dragon Bravo Fire at Grand Canyon's North Rim burns 113 structures on over 146,000 acres, destroying the Grand Canyon Lodge. One firefighter is killed.

Glossary

aerial ignition: the use of aircraft to drop incendiary materials or devices to ignite fuels on the ground.

arson: the deliberate or malicious setting of a fire with the intent to damage or defraud.

backfire: a deliberately set fire used to consume fuels in the path of a wildfire. The backfire can also be set to change the direction of a wildfire.

Bambi bucket: the device lowered from a helicopter to dip water from a pool, lake, or other source to suppress a fire.

buffer zone: an area, such as a highway, that is designed to keep a wildfire away from structures or commercial structures.

burn boss: the specialized trained professional responsible for the management of a prescribed fire.

cache: the storage or supply of tools and devices used by firefighters in a location where the items can be accessed separately from the ones carried by personnel.

complex: a designation for a wildfire when two or more individual fires or incidents combine, then assigned to a single unit of command.

containment: the point, usually expressed as a percentage, when a wildfire has a fuel break completely surrounding it. The fuel break may be natural, such as a stream, or human or mechanically constructed.

crown fire: a type of fire where the blaze moves through the crown of a tree independent of the fire on the ground.

cultural burning: the intentional, controlled lighting of fires that merges contemporary wildfire science with purposeful indigenous practices validated over generations and transmitted in oral traditions.

defensible space: the area surrounding a structure or other resource that has been cleared and acts as a barrier between a wildfire and the resource. This is usually defined as a space 30 feet or more from a structure.

demonization of fire: the belief that all fire is dangerous and must be excluded or suppressed, representing the extreme fear of fire.

drip torch: the device used by a firefighter to drop flaming liquid fuel onto an area that is deliberately being burned.

entrapment: the situation in which fire personnel are caught unexpectedly in a life-threatening place or incident in which there are no escape routes available.

extreme fire behavior: a term used to describe wildfire behavior that is unexpected because of the rate of its spread, the presence of a convection column, crowning, spotting, or other action that is erratic or dangerous.

fast fires: blazes that thrust embers into the air ahead of encroaching flames.

Federal Responsibility Area: areas where the federal government has the primary financial responsibility for preventing and suppressing fires. The Department of Agriculture, Forest Service, the Department of Interior, Bureau of Land Management, National Parks Service, US Fish and Wildlife Service, and Bureau of Indian Affairs generally protect these lands.

fingers: elements of a fire that project out from the main body of the blaze, long, narrow, and similar to the fingers of a human hand.

fire perimeter: the boundary or border of a wildfire.

fire triangle: the geometric explanation for the three elements required for combustion, composed of oxygen, fuels, and heat.

fireline: the part of a wildfire's containment or control line that has been scraped down to mineral soil.

flanks: the edges of a wildfire's perimeter that are parallel to the direction in which it is spreading.

flare-up: the point when a fire suddenly accelerates or intensifies.

flash fuels: fine materials, such as pine needles, grasses, moss, or leaves that can ignite quickly when dry.

fuel: the material that can lead to combustion and feed a fire.

ground fires: high-intensity fires that often create their own weather conditions, destroying all vegetation down to the soil level.

hand line: a fireline that is created by firefighters using tools such as a shovel or pulaski.

hardening: actions that individuals, communities, and businesses can take to reduce the ignitability of structures or other resources, such as banning the use of wood roofs or requiring non-combustible siding to make buildings fire resistant.

hazardous fuel reduction: the site-specific process of eliminating and removing materials that can ignite a wildfire, whether by hand or mechanically.

hotshot crew: the designation of a trained fire crew whose primary responsibility is to build firelines by hand.

initial attack: the personnel and equipment used after a wildfire is first discovered to protect structures and lives or other resources so that the fire does not spread further.

ladder fuel: materials that enable a fire to climb vertically from the surface to the crown of a tree or shrub.

large fire: designation referring to a blaze that burns 100 acres of timber or 300 acres of grassland.

light burning: the practice of using fire to reduce the buildup of vegetation to avoid a wildfire, sometimes referred to as Paiute forestry or the Indian way.

mitigation: the steps taken to reduce the risk or the impact of a wildfire, such as requiring vegetation management.

prescribed fire: a fire that is ignited under specific predetermined conditions to remove hazardous fuels or to improve habitat with a written and approved management plan and personnel.

reburn: the process or area in which a fire returns to burn more than once because the fuels and conditions reoccur.

slash: the material left over after an area has been logged, thinned, or cut up, including branches, tree bark, logs, and stumps.

smokejumper: the designation of a firefighter who parachutes from an aircraft to the fire area.

spot fire: a fire that ignites outside the perimeter of the main fire due to sparks or embers.

suppression: the process of containing or extinguishing a fire.

surface fires: low-intensity fires that burn rapidly, usually along the ground or in vegetation.

threatened: the critical infrastructure, such as roads, or homes and commercial buildings that are in the vicinity of a wildfire.

torching: the flare-up of a fire from the surface or bottom of a tree toward the top.

underburn: a type of fire that burns fuels on the surface but not trees or shrubs.

unmanned aerial systems: also known as drones, these craft are increasingly being used to provide ignition support for prescribed fire and wildland fire settings.

wet line: the creation of a surface barrier using water or a retardant to temporarily control the spread of a fire.

whirl: the tornado-like column of gases and heated air that carries smoke, flames, and debris into the air.

wildland-urban interface: an undeveloped area adjacent to the built environment, such as a housing development that abuts a forest or chaparral.

Index

About the Author

Dr. Jacqueline Vaughn is Professor Emerita at Northern Arizona University, where she taught in the Department of Politics and International Affairs and in the university's School of Earth and Sustainability. She holds a PhD in Political Science from the University of California, Berkeley, an MA in Political Science from San Jose State University, and a BS in Education from Boston University. She is the author of more than fifteen books dealing with environmental issues and policy. Most recently, she published *Water in the West* for Bloomsbury Publishing.